Lesbians in Television and Text after the Millennium

LESBIANS IN TELEVISION AND TEXT AFTER THE MILLENNIUM

Rebecca Beirne

palgrave
macmillan

LESBIANS IN TELEVISION AND TEXT AFTER THE MILLENNIUM
Copyright © Rebecca Beirne, 2008.

All rights reserved.

First published in 2008 by PALGRAVE MACMILLAN® in the US—a division of St. Martin's Press LLC, 175 Fifth Avenue, New York, NY 10010.

Where this book is distributed in the UK, Europe and the rest of the world, this is by Palgrave Macmillan, a division of Macmillan Publishers Limited, registered in England, company number 785998, of Houndmills, Basingstoke, Hampshire RG21 6XS.

Palgrave Macmillan is the global academic imprint of the above companies and has companies and representatives throughout the world.

Palgrave® and Macmillan® are registered trademarks in the United States, the United Kingdom, Europe and other countries.

ISBN-13: 978-0-230-60674-6
ISBN-10: 0-230-60674-1

Library of Congress Cataloging-in-Publication Data

Lesbians in television and text after the millennium / Rebecca Beirne
    p. cm.
  Includes bibliographical references and index.
  ISBN 0-230-60674-1 (alk. paper)
  1. Lesbianism on television. I. Title.

PN1992.8.L47B45 2008
791.45'6526643–dc22   2008003365

A catalogue record of the book is available from the British Library.

Design by Scribe Inc.

First edition: September 2008

10 9 8 7 6 5 4 3 2 1

Printed in the United States of America.

For Samar

Contents

List of Figures		ix
Acknowledgments		xi
1	Introduction	1
2	Image, Sex, and Politics: Cultural, Political, and Theoretical Contexts	21
3	Two Babies, a Wedding, and a Man: *Queer as Folk* and "The Lesbians"	61
4	Recycling *The L Word*	95
5	Dressing Up, Strapping On, and Stripping Off: Contemporary Lesbian Pornographic Cultural Production	135
6	*Dykes to Watch Out For* and the Lesbian Landscape	167
7	Conclusion	191
Notes		195
Works Cited		211
Index		231

Figures

2.1 From "The Misanthrope" (#351, *Dykes and Sundry Other* 37)   31

6.1 "Notes on Camp" (#368, *Dykes and Sundry Other* 70–71)   180

6.2 From "Get Me to the Clerk on Time" (#456, *Invasion of the Dykes* 83)   186

6.3 From "Intro to Queer Theory" (#399, *Invasion of the Dykes* 9)   189

Acknowledgments

This book is dedicated to Samar Habib for her faith in me and this book, for her encouragement, for her endless willingness to act as a soundingboard for my ideas, and for everything else that she brings to my life. Penny Gay read this work many times in its early stages and offered a tireless and efficient editorial eye that was very much appreciated. Thank you to Farideh Koohi-Kamali, Brigitte Shull, Kristy Lilas and Daniel Constantino from Palgrave Macmillan for seeing the potential in this text and for their work toward its publication. A significant thanks must also go to the anonymous Palgrave reviewer who offered some excellent suggestions for final revisions. To my family, Keara, Lara, and Faye, thank you for your support and encouragement. For their reviews and suggestions on an earlier version of this work, thank you to Dana Heller, Patricia White, and Julia Erhart. Thanks also to all those writers who have inspired me during the writing of this book (in no particular order): Sue-Ellen Case, Kathleen Martindale, Biddy Martin, Lisa Walker, Tamsin Wilton, Alison Bechdel, and many others. Thank you to Alison Bechdel for creating *Dykes to Watch Out For* and for kindly allowing me to reproduce images from it in this book. Thank you to Glita Supernova and Sex Intents for discussing *Gurlesque* with me and for creating this unique cultural product. Thank you to all the volunteers at the Lesbian Herstory Archives in Brooklyn for building and maintaining this great cultural resource and facilitating my access to it. Thank you to James Keller, Leslie Stratyner, Angela Ndalianis, Samar Habib, Janet McCabe, and Kim Akass for permission to reprint and for giving this work its first public airings.

Images in this book have been reprinted with permission from Alyson Books.

Portions of this book have previously appeared in the following books and journals:

"Embattled Sex: Rise of the Right and Victory of the Queer in Queer as Folk." *The New Queer Aesthetic on Television: Essays on Recent Programming.* Ed. James Keller and Leslie Stratyner. Jefferson, NC: McFarland Press, 2006. 43–58.

"Fashioning The L Word." *Nebula: A Journal of Multidisciplinary Scholarship* 3.4 (2006): 1–37. <http://www.nobleworld.biz/images/Beirne.pdf>.

"Dirty Lesbian Pictures: Art and Pornography in The L Word." *Critical Studies in Television* 2.1 (2007): 90–101.

"Lesbian Pulp Television: Torment, Trauma, and Transformations in The L Word." Refractory: *A Journal of Entertainment Media* 11 (2007). <http://www.refractory.unimelb.edu.au/journalissues/vol11/Beirne.html>.

"Mapping Lesbian Desire in Queer as Folk." Televising Queer Women: A Reader. Ed. Rebecca Beirne. New York: Palgrave Macmillan, 2007. 99–107.

Chapter 1

Introduction

This monograph began as an examination of the recent increase in lesbian images in popular culture and particularly in television. The presence of these images appeared to mark a defining moment in lesbian history, a tipping point perhaps, where the inclusion of some lesbians in the mainstream, the assimilationist trajectory of current lesbian culture, capitalism, and various other factors would produce a mainstreamed lesbianism. As I was examining these images and narratives as they were occurring or shortly thereafter, little critical material was available that discussed these representations, so I found myself utilizing critical material and theory that talked about popular cultural products from earlier times. This approach yielded fruit, as, contrary to what I had imagined, I found that this critical material was indeed highly relevant, demonstrating that my own initial supposition that everything had changed, which had also become a popular assertion in the mainstream media by that time, was at least somewhat inaccurate. Indeed, doing archival research on the media hype surrounding what was known as "lesbian chic" in the early 1990s as a grounding for my research showed me that the types of articles discussing lesbian chic in the 1990s were almost identical to those that were being written after the millennium, and these latter articles interestingly did not acknowledge that the same arguments they were putting forward had already been written ten years earlier.

The season two revamp of *The L Word*'s theme song rewrites the opening lines of *The Sound of Music*'s "My Favorite Things" as "girls in tight dresses / who drag with moustaches," and the theme song's chorus asserts that "this is the way / it's the way that we live / . . . and love" (Betty). These campy lyrics are significant in that they are emblematic of the manner in which lesbians are culturally codified in the early part of the noughties (the first decade following the year 2000), and the supposition that the way lesbians "live / and love" can be adequately expressed by a soapy melodrama is indicative of a popularized urge to define and *show* ourselves to the mainstream (always with the best possible lighting). The same aesthetic can be seen in earlier promotion of *The L Word*, most notably on the cover and in two lesbian-centered articles in *New York* magazine (January 12, 2004), which presents the "new" lesbian as coming in two shades: "lipstick" and "boi." The cover article's title, emblazoned across the more risqué sections of the naked (mainly heterosexual and conventionally beautiful and feminine) actresses' bodies, announces that these are "Not Your Mother's Lesbians" (Bolonik).[1] The rigorous and seemingly inevitable positioning of these remodelings of femme and butch styles as oppositional to an earlier lesbianism can be seen throughout contemporary discourse—from mass market examples to subcultural magazines and even in academic discussion.[2] While the protagonists shift, the oppositional and often extremely hostile outlook remains: it is almost as if any given contemporary discourse cannot conceptualize lesbians as not inevitably pitted against one another, whether for media attention or for status as most transgressive, most fashionable, or most useful to the achievement of lesbian rights.

What is lacking from such accounts is both a nuanced analysis of the multiple potentialities such images and representations may generate and the acknowledgment of lesbian histories, which leads to a tendency to "recreate" lesbianism at each turn. Such understandings fail to sufficiently elucidate contemporary lesbian subjects and representations with the diverse manifestations of these subjects and representations isolated into descriptions of the "new" sex-radical, the "new" glamorous lipstick lesbian, or perhaps the "new" boi king. Lesbianism is thereby pixilated as complex patterns of lesbian subjectivities, cultural production, and representations are broken down into individual pixels, with these tiny elements of the picture then being used to explain lesbianism.

By focusing only on television, I felt that I too was effacing the variety and multiplicity of lesbian cultural products. As Martha Gever has observed, "lesbian identities and practices will continue to take forms not yet permissible or, perhaps, even imaginable within a culture defined as fundamentally heterosexual" (44), and they will particularly do so in forums that are unhampered by interior or exterior censorships of lesbian representation, that is, those forums that are not attempting to relate to a heterosexual marketplace. I also felt that I was once again enacting a process of splitting the realms of lesbian cultural production and privileging what is seen as the "New!! Exciting!!" world of lesbian representation on television. If one examines "new" representations (the most notable of which are televisual) in isolation from other cultural products with lesbian creators, themes, and characters, one loses both perspective and a wealth of knowledge of trends and cultures, and it becomes easy to find oneself creating a narrative of lesbian cultural representation as only really existing after Ellen Degeneres, where the lipstick lesbians of lesbian chic are the first dykes to enthusiastically don a dress. Analyzing a variety of subcultural and mainstream texts that could loosely be described as part of entertainment culture, this monograph concerns itself with an attempt to contextualize and account for thematic and ideological shifts within these texts that reflect or represent lesbian subjects and cultures. It outlines the influences of the past upon the present and attempts to locate some common facets of the construction of lesbianism in the new millennium, including where "newness" does indeed reveal itself. A thorough examination of each of these texts reveals complex, unique, and often surprising positionings toward cultural discourse on lesbianism (in the form of other texts, cultural theorizing and commentary, and historical movements).

INFLUENCES

Much has been written about the impact of the sex wars upon lesbian culture, and perhaps now, an entire generation after the beginning thereof, less emphasis should be placed upon them. As Kathleen Martindale asserts,

> Since the sex wars were largely fought by (white, American) intellectuals over books and ideas and then were rehashed in more books and scholarly articles, the whole affair might seem like a tempest in a teapot. Perhaps, as it recedes from view, that will be the way feminist history will regard it. Looked at from the vantage point of another decade, however, this highly cerebral and ironically disembodied struggle set the terms and the agenda for contemporary feminist and lesbian discourses on sexuality and sexual representation. (7)

The influences of these terms and this agenda can still be seen in a multiplicity of texts emerging or continuing in the years after the year 2000. Although sex radicalism and queerness were often posed as intrinsically oppositional to commodity cultures, the stylish nature and emphasis on flexibility within these movements ironically appears to have been one of many elements that has allowed the increasing commodification of lesbianism. Reading *Gurlesque*, a lesbian stripshow, and *Dykes to Watch Out For*, a lesbian-feminist comic strip, in the context of sex wars debates, I was surprised to discover the extent to which both texts reflected and embraced aspects of both sides of the debate. This can be seen as a reunification of these, at times apparently insurmountable, rifts or an indication that cultures/texts were always more multiple than accounts would wish to suggest—a reminder not to overemphasize the gap between the positions of "queer sex-radical" and "lesbian/feminist" and presume that an ostensibly lesbian-feminist comic cannot be sex-radical and a sex-radical lesbian stripshow cannot be feminist.

The significance of 1990s queer theory to contemporary lesbian culture also cannot be underestimated. Of course, any analysis of theory's influence on culture lands one squarely in a chicken-and-egg conundrum—theory, usually reflective of trends in culture, has a hand in creating trends within a (sub)culture, which is then further analyzed by theory, and so forth. Lesbian culture seems uniquely tied up with ideas from the early seventies until today. While many within contemporary lesbian culture are unlikely to have heard of Judith Butler's notions regarding gender performativity, for example, few would be able to avoid having witnessed cultural products firmly informed by such theories, the contemporary drag king movement being perhaps the most pressing example.[3] And new forms of activism and cultural debate—around issues of marriage, parenting, the relationships between lesbian cultures and mainstream cultures, and growing amounts of gay conservative rhetoric—are reflective of and help shape contemporary cultural attitudes and products.

Several writers have engaged with the overdramatization of the conflicts of the sex wars seen in many texts, which has led to a certain demonization and renunciation of lesbian-feminism and feminism generally. Just as the lesbian characters of *The L Word* have been placed in contrast to "your mother's lesbians," in these narratives of genesis, contemporary lesbianism is often pitted against its foremothers. Biddy Martin locates this as a tendency to "construct 'queerness' as a vanguard position that announces its newness and advance over against an apparently superseded and now anachronistic feminism with its emphasis on gender" ("Sexualities without Genders" 104). Martindale expresses such a shift:

> The simplest version of the transformation goes like this: thesis–lesbian-feminism—antithesis—the feminist sex wars—synthesis lesbian postmodernism. . . . It's a story, a myth of origins about a generational changing of the guards, marking and almost always celebrating a shift from the alleged inclusivity of the boast that "any woman can be a lesbian" to the much more exclusive and unabashed elitism of lesbian (cultural, theoretical, and most of all, sexual) chic. The project of accessibility shifted to a more disturbing one of excessibility. (2)

In perhaps the most polemical manifestation of this critique, "From Here to Queer: Radical Feminism, Postmodernism and the Lesbian Menace," Suzanna Danuta Walters attempts to provide some definition and account of the then "new" queer politics and theory before going on to critique its inattention to material social relations, particularly that of gender. As is suggested by her title, Walters' main interest lies in assessing the significance and the weaknesses of queer theory and politics in addressing lesbian identity, culture, and concerns. Walters acknowledges that many "lesbians (including myself) have been attracted to queer theory out of frustration with a feminism that, they believe, either subsumes lesbianism under the generic category woman or poses gender as the transcendent category of difference, thus making cross-gender gay alliances problematic" (843). She also, however, expresses her fears both that "many lesbians' engagement with queer theory is informed itself by a rudimentary and circumscribed (revisionist) history of feminism and gender-based theory that paints an unfair picture of feminism as rigid, homophobic and sexless" and that queerness is being theorized as somehow beyond gender, displaying little attention to "the very real and felt experience of gender" (844).

Walters' articulation of the deployment of lesbian-feminism as a theoretical "other" against which culture can be created remains a pertinent insight even today. Walters is definitely not alone in such an analysis. As Shane Phelan observes,

> Recent theory and activism has offered two seemingly opposite failures in the analysis of gender's role in the politics of sexuality. As Biddy Martin has helpfully argued, much queer theory and activism has constructed itself precisely in opposition to earlier feminist work. In this construction

feminism becomes the stodgy, moralistic parent that queers must rebel against, often replacing hegemonic culture as the primary opponent. This separation of feminist analysis from queer politics has the unintended effect of reducing queer theory's potential for critical intervention into gender as a structure of power. (197)

Such insistences upon continuing to view gender and its specificities as an integral element of lesbian theorizing are, of course, largely a reaction to the excesses of postmodernist culture and the embrace of gender as performance. While some (though certainly not all) theoretical texts since Walters' and Phelan's time have gained greater complexity in their characterization of the sex wars and feminism, in other less nuanced and more simplistic forums, particularly media discourse on lesbianism (both mass and community-specific media), the "construction of an old, bad, exclusive, policing lesbian feminism" (Walters, "From Here to Queer" 848) has continued and possibly even been heightened.

The most ambitious project of this critique, however, must surely be Dana Heller's 1997 collection *Cross-Purposes: Lesbians, Feminists and the Limits of Alliance*. Contrary to what may be suggested by the title, the anthology is not a critique of lesbian-feminism but rather an attempt to conceptualize the shifting conditions and understandings of both terms and movements. Gathering together diverse writers, including some of the luminaries of lesbian and queer studies—Teresa de Lauretis, Bonnie Zimmerman, Sue-Ellen Case, and Lillian Faderman—Heller's volume acts as "a long overdue critical intervention into the history, current condition, and evolving shape of lesbian alliances with feminisms in the United States" ("Purposes" 1). A central concept engaged by several of the essays within this work is that of generational shifts:

> Generational shifts have impacted on lesbianism's uncertain alliance with feminism, particularly as a generation of lesbians came to political awareness not through the women's movement but through groups such as Queer Nation and ACT UP, organizations that placed them in a viable post-Stonewall alliance with gay men. . . . the shift away from lesbian feminism to queer was decisively marked by sex-radicalism. (Heller, "Purposes" 12)

Now, a further generation past these times, why attempt to re-energize these debates? My project in this book is to attempt to explore what relationship a new generation of lesbian texts bear to feminist, lesbian, and queer discourses and to see to what extent previous generational issues have been absorbed, digested, and recycled as new and oppositional discourses. My lesbian generation came of age not through either feminist movements or the heyday of Queer Nation but rather in the midst of what will perhaps come to be known as the mainstreaming era, an era that also, it should be noted, includes those who are vehemently opposed to mainstreaming in all its forms. For example, at "Revelling '05: Queer Lives and Spaces," a queer conference held in Sydney, a significant number of the papers given were critiques of mainstream or commodified queer culture, and the contemporary leadership of the Sydney Gay and Lesbian Mardi Gras and *The L Word* came under particular attack. As in any era, many trends exist simultaneously—the small-scale avant-gardist forms of cultural expression energize the large-scale commercial forms with ideas and are energized by antagonism toward them.

Linda Garber takes this debate further by writing against the characterization of these debates as generationalist through situating the roots of queer theory and lesbian-feminism as inextricably intertwined and several of their central precepts as undeniably similar. In the intellectually astute *Identity Poetics: Race, Class, and the Lesbian-Feminist Roots of Queer Theory*, Garber seeks to

> demonstrate the links between two schools of thought that are frequently pitched as opponents in ideological battle—often figured as a generation gap between lesbian thinkers—and to restore to their central place in the story the writings of working-class/lesbians of color whose marginalization is foundational to the misbegotten construction of the debate itself. The debate between lesbian feminism and queer theory (or, as it is often more broadly labelled, postmodernism) presents a simplistic either/or choice between two terms that are mutually implicated. (1)

In so doing, Garber attempts to advance beyond the dualistic nature of such debates by demonstrating their similar roots and theorizations instead of reiterating a simplified popular understanding of a movement/culture without taking into consideration its complexities. As Garber has already articulated the "nuanced genealogy of lesbian and queer theories" (7), I will not reiterate these genealogies here but rather will attempt to take up what bearing these genealogies and the debates over them have had upon contemporary texts of entertainment media, cultural commentary, and academic work. The arguments of Garber, Heller, Martin, Walters and others certainly do not appear to have fully permeated the consciousness of the majority of contemporary cultural producers and commentators, and much discourse on lesbianism is still propelled by a teenage oppositionalism to the perceived maternal authority of a previous lesbian generation.

A New Generation?

This generation is perhaps the clearest generation of "nots": Ellen DeGeneres, lesbian heroine of the 1990s, took some time to adjust to the term lesbian, infamously asserting that the term "sound[s] like somebody with some kind of disease" (DeGeneres qtd. in Handy, "Roll Over" 78);[4] the ladies of *The L Word* were nudely announced on the cover of *New York Magazine* as "not your mother's lesbians"; and we are already in the midst of the second wave of "lesbian chic" that is completely new and has certainly *not* happened before.[5] Martin in 1994 noted that "[t]he popular gay press has been full of comparisons between the dull, literal-minded, up-tight seventies and the sexually ambiguous, fun, performative nineties" ("Sexualities without Genders" 104). Always two steps behind, the popular mainstream press has recently been full of comparisons between the dull, literal-minded, uptight (nonspecific) "past" and the sexy, fun, performative noughties. Certainly, this is considerably due to elements of lesbian culture now being situated within a commodity culture that must constantly recycle and relabel trends and fashions to remain viable. Case takes up the issue of commodity culture toward the end of her sardonic "Toward a Butch-Feminist Retro-Future." Indeed, Case's fear that "[w]hat was once a lesbian or gay community is now becoming a market sector" (212) appears to be confirmed by the emergence of gay and lesbian cable networks or the desire seen in gay magazines and political groups or such cultural products as *Queer as Folk* to constantly re-emphasize the significance of the gay and lesbian market and that market's power to *buy* acceptance and respect.

The resignification of "queer" as a market sector or a marketable commodity is one that eliminates all political or oppositional meaning from the term queer, and the term becomes, as Sally Munt asserts of gayness in the British *Queer as Folk*, "formulaically rebranded as attractive and aspirational, it has acquired cultural and symbolic capital, it has, through commodification, become *respectable*" ("Shame/pride dichotomies" 539). As Case points out, however, through her reading of Alexander Chee's account of Queer Nation, the element of individualism and respectability within dominant paradigms was present in the group, and thereby in the term queer, from the very beginning:

> Queer coalitions, then, in a hurry to get onto the streets, began to interrogate the "normal," as if outside of the normalizing operations of patriarchy, capital and nation. For, without arduous attention to dominant contexts, single-issue politics operate within them. The term *queer*, then, circulating out from Queer Nation, asserts itself as an umbrella term without the hard rain of coalition-building. Thus it reinstates the dominant social structures, lending its power to those who are already vested in the system, with the exception of their political identification. Not surprisingly, then, white middle-class men will form the constituency. Their culture, sub or not, will continue to be representative. ("Towards a Butch-Feminist Retro-Future" 218)

According to such an analysis, perhaps the divesting of all identities except gay white men from the meaning of "queer" in *Queer as Folk* (a thorough discussion of which is undertaken in Chapter 3) is not so much a resignification—perhaps it was there all the time. I must, however, demur at the implication that "queer" was not and is not a term invested with multiplicity, with at least an idea and desire for inclusivity and openness. While many of the hopes for "queer" may now have been dashed, the extent to which many individuals and groups, including Case herself, were taken by the possibilities of the word "queer" suggests that even if the possibilities of such exclusionary definition were present in the term from its reclamatory inception, there was also a more utopian impulse and, to some extent, practice. The current definition of "queer" seen in such texts as *Queer as Folk* is, if not an entire reversal of previous understandings of queer, at least an intensification of exclusionary classification. De Lauretis writes that the neutralization of difference seen in the deployment of queer theory is "indeed the opposite to the effect I had hoped for in coining the phrase *queer theory*, by which I wanted to displace the undifferentiated, single adjective *gay-and-lesbian* toward an understanding of sexuali*ties* in their historical, material and discursive specificities" (46).

Chapter 3 also examines the manner in which lesbians are represented as outside the sexual economy, tied to home and childbearing, while the gay male characters exist in a world of backrooms and anonymous sex. Lesbians are thereby also, according to the cultural politics of *Queer as*

Folk, situated outside potential radicalism and necessarily imbricated with assimilationist politics. Walters asserts that such characterization is common to the "new queer culture" wherein "gay male sex and *its* histories have become the very model of radical chic" ("From Here to Queer" 850), while lesbian sexual history is characterized as nonexistent and lesbians are portrayed as puritanical. While repression of certain forms of sexuality and gender expression have certainly taken place, the emphasis on the "lesbian sex police" as the ultimate repressors, instead of the mechanisms of repression enacted by heterosexist and homophobic society, is curious. Interestingly, Walters then goes on, however, to suggest that anonymous sex spaces (and presumably other lesbian sex of the "radical chic" variety) *are* in fact based upon male models: "Are lesbians unable to construct, envision, imagine, enact radical sexualities without relying so fundamentally on male paradigms?" (850). A corrective to this notion can be seen in an anecdote of Case's:

> Slapping each other on the back, we joke, "was lesbian s/m invented by Gayle Rubin and Pat Califa in an argument with antiporn advocates?" Leafing through our old phone books and photos of friends flamed out in one affair after another, we snorted at the queer dykes' belief that they were originating the practice of multiple sexual partners, s/m scenarios, the use of sex toys, and the habit of hanging around bars.... What a surprise then, to learn that queer dykes associated such sexual promiscuity as more narrowly particular to a gay male culture that they would then need to assimilate and imitate. ("Toward a Butch-Feminist Retro-Future" 209–10)

Whether contemporary lesbian sexual cultures emerged from gay male cultures or a prior lesbian culture, the current popularity of masculinity studies both within and outside of queer analysis cannot be denied. Of course the most visible and pronounced element thereof of relevance to this monograph" is the contemporary drag king movement—which includes a significant discourse within the academy—and this popularity is reflected in some engagement with drag kings in each of the texts I examine.[6] Female masculinity in its many forms has been both celebrated and disdained, often simultaneously, in lesbian communities and cultural productions.

Lesbian Genders

Manifestations of female masculinity have long been targeted for "normalization" by various elements of both lesbian communities and those who attempt to depict them. Early examples include the Daughters of Bilitis, who encouraged their members to don frocks to appear more acceptable, and this tradition continues with lobbyist organizations such as the Human Rights Campaign, who attempt to project feminine images (Phelan), and with contemporary television shows such as *The L Word*. Increasing mainstream lesbian representation in the 1990s was dominated by feminine or at the very least *feminized* images, and even lesbian film often hesitated to depict truly masculine women.[7] Even the seeming celebration of female masculinity seen in the drag king scene or boi cultures can at times be posed as oppositional to other manifestations of female masculinity, particularly butchness.[8] These representations interestingly occurred concurrently with a celebration of butchness within other elements of lesbian cultures and academic theory. The 1990s saw the emergence of multiple academic articles and texts that theorized butchness or discussed butch women's position in various cultures.[9] Works by British painter Sadie Lee, such as "Raging Bull," "La Butch en Chemise," and "Venus Envy," or the photography of Della Grace (now Del La Grace Volcano) powerfully represented visibly butch subjects, and though femme lesbians and lipstick lesbians were beginning to assert themselves, lesbian cultures more generally were (and, I argue, still are) dominated by an alignment of lesbian authenticity with lesbian masculinity. Interestingly, the 1990s *Vogue* "drag king" spreads that Lewis and Rolley presented (181) are in many respects very similar to the contemporary drag king and butch photography seen in the current crop of lesbian magazines, perhaps suggestive of not so much a shift but rather a continuation of a tradition that cycled back into fashion. In this respect, the current theoretical blossoming of discussion of drag kinging can be seen as an extension of the popular discourses surrounding butch subjectivity and drag queening in the 1990s, though there are definite ideological distinctions between these discourses.

Female masculinity, even in texts that do not visibly display it, is presented as inexorably enmeshed with true lesbianism. And, somewhat curiously, the wealth of texts, both academic and experiential, that examine femme lesbians and attempt to place them as authentic lesbian subjects or to account for them within lesbian history and literature (whose writers include but are not limited to Joan Nestle, Amber Hollibaugh, Cherríe Moraga, and Lisa Walker) have remained largely separate from discussions of "lesbian chic." None of the accounts or analyses of lesbian chic I have read have, for example, taken up Danae Clark's cheeky parenthesized supposition in "Commodity Lesbianism" that "(I can't help but think, for example, that the fashion controversy may not be about 'fashion' at all but has more to do with the fact that it is the femmes who are finally asserting themselves.)" (199). This inattention may be due to perceptions that the "lipstick lesbians" seen in mainstream culture are not femmes, or discomfort at applying subculture specific methods of analysis to mainstream products, or the sensation that the femininity of these characters is a gesture of assimilation rather than desire or identity-assertion. However, despite these potentialities, some attention to the understandings of feminine lesbians elucidated by these works would be pertinent and enriching to discussions of contemporary examples of lesbian femininity in the mainstream and may assist in preventing the dismissal of femininity in lesbians as automatically pandering to heterosexuality. Although in many "lesbian chic" texts, including *The L Word*, that is indeed what it appears to be, it is unnecessary to dismiss the history, significance, and, most importantly, the authenticity of feminine lesbians in the name of a critique of mainstreaming. It could be said that in some ways the celebration of gender subversion and gender crossing has had the ironic effect of reinforcing normative perceptions of gender by reinforcing the masculine (female) subject as the active/desiring/authentic subject, leaving the feminine (female) subject as the passive/desired/performing object. This issue will be further discussed in Chapter 5 in relation to lesbian pornography and the critical perception that it is only through embracing the butch role that performers can become subjects rather than objects of the gaze.

The abnegation of lesbian femininity often coexists with an abnegation of the female body. De Lauretis critiques the

> new imaginary [that] has developed out of the progressive repudiation of the female body: the discourse on sexuality has moved from the impossibility of a feminine identity theorized by feminists since the late 1970s, to the alleged "subversion" of gender identity in queer/lesbian studies. (47)

Such a contention was earlier made in "Sexualities without Genders and Other Queer Utopias" by Martin, who conceived that: "antifoundationalist celebrations of queerness rely on their own projections of fixity, constraint, or subjection onto a fixed ground, often onto feminism or the female body, in relation to which queer sexualities become figural, performative, playful, and fun" ("Sexualities without Genders" 104). Martin further notes: "a resistance to something called 'the feminine,' played straight, and in a tendency to assume that when it is not camped up or disavowed, it constitutes a capitulation, a swamp, something maternal, ensnared and ensnaring" (105). This resistance to the feminine Martin identifies is of central importance to *Lesbians in Television and Text after the Millennium* because this resistance is unexpectedly identifiable in many contemporary texts that are themselves the products of "lesbian chic," which is commonly understood as celebrating femininity. While both *Queer as Folk* and *The L Word* largely present what could be called "lipstick lesbians" and overtly disparage nonfeminine lesbians, they share a view of feminine lesbians reminiscent of representations of the "feminine invert" and evocative of similar depictions seen in pulp novels of the 1950s and 1960s.[10] These contemporary characters are indeed presented as if through their femininity they are performing a capitulation, a gesture of assimilation, and the authenticity of their lesbianism is questioned. Due to the highly noticeable and widely critiqued lack of nonfeminine visual representation in these programs, an assertion that their presentation of lesbian femininity is not uniformly positive may appear to be very strange. However, textually there *is* a high level of tension between text and subtext in these series, and what is ultimately signified is not necessarily what one would expect.

In a 1993 issue of *Signs*, Walker took up the question of how the figure of the femme is addressed in some contemporary works of feminist/lesbian/queer theory. In her reading, Walker identifies a certain "privileging of the visible" in the construction of lesbian subjectivities, identifying this privileging as being instrumental to the elision of "other identities that are not constructed as visible . . . with the result that these identities remain unexamined" (874). Walker argues that within the schema of Case's "Towards a Butch-Femme Aesthetic," "the femme is invisible as a lesbian unless she is playing to a butch" (881), and this invisibility renders her inarticulable and inauthentic as a lesbian subject. Walker also observes a similar maneuver in Butler's work, as

> [t]he femme's radical desire is understood to offset her normative external appearance. . . . [However] the femme cannot be conceptualized except as part of a unit in which her desire is constructed as internal to the butch's external expression of that desire. Butler's expectation of coherence between (outer) sexual style and (inner) sexual consciousness puts her back within the framework of culturally intelligible identities that she seeks to destabilize. In the process of illustrating how the sexual styles of drag queens and butches visually denaturalize the categories of sex and gender by producing signs of their nonconformity to heterosexual norms on the body, Butler

establishes them as coherent subjects in terms of political rather than gender intelligibility. That is to say, radical consciousness is ascribed to radical appearance. The flip side of this equation is that orthodox ("straight"?) consciousness is ascribed to conventional appearance. (885)

The perception that without a butch, the femme is "de-lesbianize[d]" (Ciasullo 599) and that a more orthodox appearance connotes depoliticization is also endemic in academic and cultural discourse around feminine lesbians in the latter years of the twentieth and early years of the twenty-first century, particularly within discussions of mainstream texts. While in visual texts it is obviously easier to signify a character as "lesbian" by utilizing a masculine appearance, and indeed, real-life lesbians are most easily visually identifiable through markers of masculinity, surely at least in our theoretical discussions we should be able to conceptualize desire and politics outside of the realms of external visual representation.

Indeed, critique of lesbian mainstreaming is often framed through a critique of lesbian femininity or at least a critique that situates femininity as causally rendering lesbians invisible or commodifiable, desiring to pass or otherwise uncomfortable with their lesbianism. Lisa Pottie writes that

> the variations on femme make it easier for lesbians to not be seen, or if seen, to be reassuringly presented in the media as girls "who just wanna have fun" . . . less acknowledged in the assumption of the subversiveness of femme identity is the political reality of the continuing pressure to conform and the desirability of passing in many circumstances . . . Feminine appearing lesbians put pressure on other lesbians to conform, either directly or indirectly by ignoring them on the cruise. (n.p.)

Such views are perhaps most clearly extrapolated by Ann Ciasullo, who in her analysis of "Cultural Representations of Lesbianism and the Lesbian Body in the 1990s" suggests that "[p]erhaps the configurations of single and coupled femmes work to undo the 'lesbian' signifier and to de-lesbianize the subject for mainstream audiences" (599). The situation is perhaps more complicated than such accounts suggest, however, for while it is indeed the case that feminine lesbians are most clearly and visibly represented in mainstream and are increasingly represented in subcultural depictions, such *visibility* is not necessarily connotative of a celebration or even a reinforcement of lesbian femininity even in the most apparently femme texts. Pottie uses "lesbian chic" as evidence that Walker's discussion of overvaluing visibility and conflating it with authenticity, which renders the butch as the authentic lesbian, no longer stands in late 1990s culture. The connections between visibility and authenticity and the inability to articulate the femme that Walker locates in various 1990s queer theorists in her 1993 essay do indeed still hold true more than a decade later in both theoretical dialogue and some of the mainstream texts that Pottie would now perhaps see as exemplary of their repudiation. In both *Queer as Folk* and *The L Word*, the masculine lesbian is figurally all but absent but is subtextually coded as the "real lesbian," while the feminine lesbian is simultaneously hypervisible (though arguably the camp femme is not) and yet signified as "not real." What seems to be emerging herein are two seemingly counterposed forms of gender policing—lesbian masculinity remains unseen, while lesbian femininity remains unreal.

Theorists viewing mainstream culture tend to be particularly quick to dismiss *or* celebrate feminine representations of lesbianism in terms of capitulation; that is, visual proximity to heterosexuality is often viewed as either negative when it is seen to be part of a fantasy rendition of lesbians as consumable (Ciasullo) or positive if it is seen to upset perceived boundaries between homosexuality and heterosexuality (Inness). While great variation exists within such responses, I am troubled by the tendency to always view femininity in relation to heterosexuality. Perhaps, however, such analyses are simply a matter of audience; the possibilities of lesbian femininity seem more fragile when placed into a heterosexist capitalist economy, and the critics understandably become more jittery when the gaze is intended to not only be a lesbian one. Analysis of the added ramifications of lesbianism entering into mainstream consumerist society is not new. The much-vaunted appearance of "lesbian chic" in the late 1980s and early to mid 1990s produced a spate of both polemical and critical perspectives on the phenomenon, often addressing the potential impact of mainstream assimilation and voyeurism.[11] Here the majority of the focus was upon print texts—the iconic *Vanity Fair* and *Newsweek* covers as well as lesbian stylings in fashion magazines and various other magazine and newspaper articles—although attention was also paid to various lesbian celebrities of the era and filmic depictions of lesbianism.

Perhaps the most famous reading of "lesbian chic" is Clark's "Commodity Lesbianism," which provides a very engaged and contemporary reading of the phenomenon. Later readings of the period, such as those undertaken by Ciasullo, Inness, and to a lesser extent Gever, share a certain ambivalence about the ability of the lesbian images present in that era to have a positive effect in fighting homophobia but see this more as being due to the feminized nature of such images. It is interesting that readings of lesbian chic such as those undertaken by Clark or Lewis and Rolley focus on the often *masculine* images in fashion of the period, while later writers are almost always engaged with the *femininity* of the phenomena. Perhaps this is indicative of a reading of lesbian chic through the lens of a certain retrospective perception shaped by the increasingly feminized images of lesbians in the late 1990s, or perhaps it is simply a coincidence. Certainly, however, it would seem to suggest that there was more than one gendered dimension to "lesbian chic."

These later readings of the period are often framed by a discourse of "positive or negative images" that is endemic to analysis of lesbian representation in mainstream texts. Therefore, the complexities in understanding the gaze introduced by both the sex-wars–inspired theorization of butch-femme and pornography and queer theory's emphasis on performance are glossed over or sometimes completely ignored. While the discussion of femme lesbian identities that emerged during the sex wars and the feminine representations and "lipstick lesbians" of lesbian chic may have opened some subjectival and representational space for feminine lesbianism, the figure of lesbian femininity is still very constricted—either fetishized or disparaged, with both arguments taking place only on the surfaces, and images, of discourse.

Discussing arguments that see a direct causal link between pornography and violence against women, Maureen Engel notes a "conflation of reality and representation" (102). One can also witness such a conflation in many texts that discuss mainstream lesbian representations, especially those on television. It is difficult not to feel that homophobic representations of murderous lesbians, such as those still seen on such programs as *Law and Order* (which has included two recent episodes in which women have murdered their female lovers),[12] must have some negative influences on perceptions of lesbians in dominant culture. Despite the occasional continuation of such story lines, dominant in early televisual representations of lesbians, there have certainly been qualitative and quantitative improvements to the representations of lesbians on television. This is in large part due to those activists and cultural producers who have promoted the importance of having positive images of gays and lesbians in the mainstream media.[13] Unfortunately, the discourse of positive and negative images that thereby arose has seeped into discussion of more nuanced portrayals, which are often also critiqued under the same rubric: negative representation equals harm. What makes such discussions chaotic as well as limiting is that in cases such as *Buffy the Vampire Slayer*'s Willow (Alyson Hannigan), *Queer as Folk*'s Melanie (Michelle Clunie) and Lindsay (Thea Gill), and even *The L Word*'s many lesbians, there is much, and often diametric, disagreement as to which aspects are negative representations and which are positive. More often than not, such arguments are framed by recourse to the term "stereotype." Again, there is widespread disagreement as to what exactly "the" stereotype, as it is frequently expressed, is (the use of stereotype as a pejorative is further discussed in Chapter 2).

Gazing at Lesbian Cultures

The political implications of the gaze have been the subject of much discussion in feminist theory. Laura Mulvey's 1975 essay "Visual Pleasure and Narrative Cinema" has become an almost canonical text among feminist theorists who engage in the discourse surrounding women's visual representation in many mediums. In analyzing various texts, attention must also be paid to the further lesbian critiques of the heterosexual (usually male) gaze and to the charges of male identification thereby laid at the feet of many lesbians who have been seen as objectifying women. This is then complicated by the counter-discourse surrounding objectification that arose during the sex wars, particularly in discussions of pornography and femme subjectivity. As Mary T. Conway has put it, "Gendered spectator theories have been complicated beyond Laura Mulvey's seminal essay" ("Spectatorship in Lesbian Porn" 92). Many negative reviews and critiques of *The L Word* have focused upon its objectification of lesbians: since heterosexual men are to some extent a target audience of the glamorous, feminine, and commodified lesbians represented in the series. This has led many to proclaim *The L Word* to be "pud fodder for Joe Sixpack" (McCroy). It is no coincidence that at a recent academic conference panel on *The L Word*,[14] Mulvey was one of the most cited theorists, as the commodified and more mainstream status of *The L Word* reintroduces the specter of the male viewer. The second season of *The L Word* complicates these discourses by presenting story lines that examine the gaze, implicating viewers of all genders and orientations (but particularly heterosexual men) in voyeurism (see Chapter 4).

What are the impacts that products like *The L Word* have on lesbian culture? The sheer scope of the trend toward commodified images, together with the large amount of attention they have garnered, seems to have led to a certain pessimism on the part of those who champion more oppositional or marginal(ized) cultural practices. The feeling is often that the two cannot be simultaneously in ascendancy. Engel, for example, in her theorization of lesbian pornography, sees the advent of lesbian chic in the early 1990s as heralding the death knell of lesbian pornographic, or indeed lesbian subcultural, production.

> So how does participation in this new commodity market undermine, or at least neutralize, the oppositional politics of lesbian pornographic production? The answer lies in the fact that lesbian pornographic productions "fail" to become commodities, at a moment where when North American culture at large is turning lesbians themselves into exactly that. As a result, lesbian pornographic productions seem more and more in the late 1990s to represent not interesting and contradictory forms of cultural critique and community building, but rather poor imitations of their more affluent gay and straight bedfellows. They begin to lose their allure and cultural influence when they are put in direct competition with the bright and shiny packages which can't help but seem so much "better" in a world which idolizes exactly those qualities (220).

Engel's anxieties for the future of lesbian porn are well founded. And yet the current state of lesbian cultural production is neither as homogenous nor as assimilationist as the celebratory narrative of Kera Bolonik's "Not Your Mother's Lesbians" or the gloomy one of Engel would suggest. The persistence and tenacity of oppositional lesbian cultural production is remarkable and promising; some things are simply not recuperable by heterosexual consumerist society, and some things will remain virulently ideologically opposed to it. In terms of the specific context of lesbian pornography discussed by Engel, the analysis of contemporary lesbian pornographic cultural production I undertake in Chapter 5 indicates that her fears are, if not unfounded, at least yet to eventuate.

Examples of contemporary lesbian video pornography I examine in Chapter 5 include the American films *Sugar High Glitter City* (2001) and *The Crash Pad* (2005) as well as English films *Tick Tock* (2001) and *Madam and Eve* (2003). This is certainly not an exhaustive list of lesbian pornographic texts available. Established companies such as Blush Productions/Fatale Media, the dominant name in lesbian video pornography in the late 1980s and 1990s, have continued to release titles post-2000, while new companies such as the San Francisco–based Pink and White (producers of *The Crash Pad*) and the Australian Bizante Films (see Katrina Fox 12) have been established. *The Crash Pad* performs a surprisingly complex examination of gender and sexual subjectivities, while magazines such as the Sydney-based *Slit*, which began publication in 2002, continue the underground, oppositional aesthetic Engel celebrates.

So perhaps the increasing amount of mainstreamed lesbian representation does not have to "undermine, or at least neutralize, the oppositional politics of lesbian pornographic production" as Engel suggests (220). Indeed, there are several ongoing contemporary examples that display "interesting and contradictory forms of cultural critique and community building" (220), engaging in both the sex-positive discourse prompted by the debates of the sex wars *and* the feminist materialist critique of sexist society, as the two are inexorably interwoven. The example directly engaged with in this study is that of Sydney-based lesbian strip club *Gurlesque*, a close analysis of which is located in Chapter 5. As Garber and others have observed, there are certainly ancestral and theoretical linkages between the discourses of lesbian-feminism and lesbian postmodernism. Nonetheless, the combination of lesbian-feminist and sex-positive principles in the lesbian stripshow *Gurlesque* is certainly surprising at first,[15] as the academic and popular discourse generally surrounding lesbian pornography appears to be driven by a sense of rebellion against a feminist "order."[16]

Likewise, texts such as the comic *Dykes to Watch Out For* (1982–), categorized by various critics as "stuck in the collectivist lesbian-feminist political vanguard which currently seems out for the count" (Martindale 58), contain *both* lesbian-feminist elements and the sex-positive discourse understood in many theories to be its arch-nemesis. Indeed, *Dykes to Watch Out For* forms the endpoint and conclusion of my monograph for precisely the feature that prompts critics Kathleen Martindale and Gabrielle Dean to critique its author, Alison Bechdel: diversity. Over its many years in print, *Dykes to Watch Out For* has documented in fictionalized form many trends and aspects of lesbian cultures. Despite the shifting trends it has documented, it has maintained a diversity of opinions: issues are discussed and argued out without characters necessarily coming to a conclusion or a definitive answer. Visually, the comic contains a butch/genderqueer sex-radical, an unreformed lesbian-feminist, luppie (lesbian yuppie) mothers, lesbian theoryheads, and a lesbian neoconservative, among others. Although these characters are frequently at odds with one another, they are presented in *dialogue* with one another, coexisting within the space of Bechdel's comic. These characters thus perform the task of not only presenting multiple cultural histories and perspectives but also metaphorically demonstrating that the currently fashionable aspect of lesbian cultures is not the only one in existence at any given time. It seems like an obvious point, yet it is one that often eludes those who perform cultural commentary.

Although I inevitably view the texts I engage with in certain perspectival, theoretical, and experiential lenses, I, like Engel, am unwilling to take an approach that "would bend the material in the service of the argument" (7) any more than is absolutely necessary. To this end I do not engage in a particular methodological or theoretical approach but rather deploy various academic and nonacademic sources and perspectives in an attempt to elucidate elements of the discourses circulating around lesbianism after the millennium. I do not primarily cite those "theorists with a great deal of institutional currency" (Garber 202), as would perhaps be expected of a text engaged with queer and feminist theories, not just because they are not necessarily the most invigorating to the study of my particular subjects but also because the very privileging of currently fashionable sources has at times led to a certain critical homogeneity. My focus on contemporary texts has created several difficulties, but these difficulties have at times been quite fruitful to the principles that underlie my research. Gathering source materials specifically pertaining to texts I am examining has obviously been difficult due to the slowness of academic publication, but it has led me to look outside the immediate discourse and at times to discover the historical underpinnings of various motifs and frames. And the ever-shifting meanings and contexts keep one mindful that culture and cultural products are not static entities able to be captured by a single discursive framework.

CHAPTER 2

IMAGE, SEX, AND POLITICS

CULTURAL, POLITICAL, AND THEORETICAL CONTEXTS

THIS CHAPTER IS INTENDED TO INTRODUCE THE KEY POLITICAL, cultural, theoretical, and textual contexts that are relevant to the study of lesbian cultural production after the millennium. The first few sections herein discuss the major cultural and political movements and issues in operation in the early years of the 2000s, while the latter sections specifically address the contexts of lesbian representation in and production of television, pornography, and comics.

MAINSTREAMING?

A major recent subject of gay and lesbian polemical works has been assimilation or mainstreaming, with ardent proponents both for and against.[1] To boil down an extremely complex issue into a simple explanation, the debate could be characterized as an argument regarding whether homosexuals (usually excluding transgendered and transsexual people and bisexual and queer people) need to simply prove to mainstream heterosexual society that they are just as "normal" as them and hence just as deserving of civil rights (a perspective often declared "assimilationist") or whether it is society itself that needs to be changed in order to accommodate sexual and gendered diversity (a perspective often characterized as "liberationist"). Although such divisions have existed within gay movements since their beginnings, the increasing presence of gay and lesbian issues and images in contemporary mainstream culture makes the threat or promise of assimilation all the more real. I will not here undertake a full explanation of the various arguments and proponents thereof in recent times, as writers such as Steven Seidman in his introduction to *Beyond the Closet: the Transformation of Gay and Lesbian Life* (2002) and Paul Robinson in *Queer Wars: The New Gay Right and Its Critics* (2005) have already done so, but I will offer a brief overview of each position.

The first argument, in Bruce Bawer's words, asks for "a place at the table" of mainstream, heterosexual society. This argument asserts that "[t]he vast majority of gays want to 'live ordinary middle class lives.' And the best path to social acceptance is to educate the public about the common human bonds between homosexuals and heterosexuals" (Seidman 3). This strategy shows that gays can lead "normal" middle class lives, pay mortgages, have families, and be as much "like" heterosexuals as possible. Because within this schema homosexuals must consistently prove to heterosexuals that they are "normal" and hence worthy of full citizenship, those who do not fall into preestablished categories of social acceptability (by being too sexual, too poor, too gender-ambiguous, too "other" to be easily recognizable as "the same") must be reformed, reprimanded, or swept under the carpet. As such, this strategy insinuates and sometimes directly asserts that there *is* something intrinsically wrong with sexualities that deviate from a nuclear-family model and that heterosexuality and its formal structures are indeed an ideal to aspire to and emulate, as seen in patronizing calls for gays to embrace "*civilized* commitment" (William Eskridge, my emphasis). This argument does not necessarily demand equal rights for those with nonnormative sexualities; rather, it demands rights simply for those homosexuals that behave with sufficient decorum to warrant it.

Gay conservatives not only "repudiate the gay movement's affiliation with the left" but also "seek to rescue homosexuality from its association with gender deviance—with effeminate men and mannish women" and "urge gays to restrain their erotic behaviour," in the words of Robinson in *Queer Wars* (2). The most notable proponents of the assimilationist position include Bawer, Andrew Sullivan, and Eskridge, together with such rapidly growing groups as the gay Log Cabin Republicans in the United States. It should be noted that the writers on the assimilationist side are usually male, white, and middle to upper class, and it is this group that has the most to gain from the achievement of such a place setting, as they are only fractionally denied it already. For those whose gender, gender performance, race, class, or sexual subculture already places them in a disadvantaged position, there is less to gain through a movement that does not take into consideration other systems of oppression. In fact, this movement frequently distances itself from those who are perceived to have the potential to stain its image.

The works of these writers sparked an outpouring of argument for "the renewal of the radical activism of gay liberationism" (Seidman 5), which sought to interrogate the structural inequities of society and build coalitions with those who also suffered from discrimination. Such writers include Urvashi Vaid, who declares that "[m]ainstreaming has been at once the goal and the logical consequence of our successes. . . . But the limits of mainstreaming are equally evident today. The liberty we have won is incomplete, conditional, and ultimately revocable" (3). In addition to Vaid, Michael Warner argued for "sexual autonomy" (1), while Richard Goldstein sought to document the "homocons," as he calls them, and to point out the "heartlessness of this enterprise and the threat it poses not just to the gay political agenda but to the ethos that links us with feminism, multiculturalism, and the entire progressive tradition" (xi). While Lisa Duggan characterizes this movement as "The New Homonormativity" and discusses the neoliberalist approach of such perspectives, Maureen Engel recalls "the warnings of Christopher Castiglia and Elizabeth Freeman, that the 'new' is not always progressive, and that an invitation to inhabit the inside rather than the outside leaves those same distinctions in place" (220). Others, such as Seidman, acknowledge that "[e]ach side has . . . both grasped and missed something fundamental about the present. Perspectives that spotlight expanding individual choice and social integration are compelling because they speak to real, dramatic developments in gay life" while still sharing "the view with those on the left side of the movement that gay inequality is deeply rooted in American society. A movement aimed at gaining equal rights and respect for a social minority leaves heterosexual dominance intact," and Seidman therefore favors "a movement that broadens its agenda beyond seeking equal rights for sexual minorities to addressing wider patterns of sexual and social injustice" (6).

It must be noted that lesbians exist in a somewhat precarious relationship to mainstreaming. During the advent of these debates, lesbians were used in various ways, though rarely were they seriously included in the discussion except in a tokenistic way. Robinson writes that "gay conservatism has been a largely male affair. The figures in the movement are nearly all men, and their writings focus heavily on the masculine case," listing as the single exception "Nora Vincent, [who] has made a career of denouncing what she considers the mindless radicalism and terminal dowdiness of lesbian intellectuals" (3). Eskridge celebrates that, as the result of "the health crises of the eighties," gay life "gained in dignity" by gaining a "lesbian-like interest in commitment": a step on the road to "self-civilization" (58), which would suggest that lesbians are indeed almost as "civilized" as heterosexuals but have not gained the privileges Eskridge associates with attaining this state of civility, nor are lesbians considered part of "gay life" in this formulation. From the other side of the debate, Michael Warner enacts a similar preconception by creating a list of possibilities for sexual outlawry—for which a dyke needs a dildo to gain entry (32). Often also subject to economic and social disadvantages as women, lesbians are less likely to gain or benefit from "a place at the table" than the frequently wealthy white men who write such treatises, and lesbian mainstream visibility has not only arrived at a somewhat slower pace but is also fraught given the problematic issue that lesbian visibility in the mainstream can often be intended and utilized for mainstream heterosexual male consumption.

The early noughties glory years of gay visibility in the mainstream, while often referencing lesbians in their mantra of "gay and lesbian," often did little more than that. In an article entitled "Vanity Fair the Latest to Use Gay Men as Lesbian Equivalent," Sarah Warn notes,

> In their rush to trumpet the dawn of a New Gay Era in order to sell more newspapers/magazines, many mainstream journalists and editors have allowed shoddy journalism to prevail, because it's easier to talk about gay men as if they represented all gay people rather than have to explain the differences between the number and ways in which gay men and women are portrayed on TV.

However, the 2004 release of the television series *The L Word* prompted a renewed interest in the "chic-ness" of lesbianism within mainstream media sources, reminiscent of a similar phenomenon approximately a decade earlier. The most significant feature of the articles that appeared during this period is that they were focused on style. The *New York Times* ran stories about "The Secret Power of Lesbian Style" (Trebay), while *Vogue*'s Irini Arakas examined elements of *The L Word*'s Katherine Moenning and her character Shane's wardrobe in order to tell her readers "What Women Want." Perhaps the most notable difference between the first blush of lesbian chic in the nineties and the second in the noughties is that the 1990s magazines that heralded "Lesbian Chic: The Bold, Brave New World of Gay Women" (*New*

York May 1993) featured lesbian celebrities such as kd lang or Melissa Etheridge on their covers, while during the second round, stories such as "TV's Gay Heat Wave" (*Vanity Fair* December 2003) and "Not Your Mother's Lesbians" (*New York* January 2004) are represented by covers featuring predominantly heterosexual actors and actresses who play homosexual characters (though sometimes the actresses don't even play homosexuals: the women featured on *Vanity Fair*'s cover are the (heterosexual) actresses from *Will & Grace* rather than those from *The L Word*).

There are certainly differences in the manner in which gay men and lesbian women are taken up by the mainstream. *Queer Eye for the Straight Guy* (2003–) showed that gays could be extremely popular to mainstream viewers, and it even spawned the spin-off series *Queer Eye for the Straight Girl* (2005–), whose cast members notably included only one lesbian. The manner in which *Queer Eye*, or even *Queer as Folk*, is marketed in contrast to *The L Word* highlights a significant difference in how gay people are positioned in relation to the mainstream; while gay men can be used to *sell products to* the mainstream, lesbians *are the product* to be sold: "Under Lesbian Chic, lesbians become at once commodities and consumers, packaged as an erotic other from the heterosexual mainstream" (Engel 219). Engel argues that this could be "the [consequence] of not fully participating in the capitalist project" (217) that the marginal and oppositional nature of earlier lesbian cultures had engendered. Despite the complicated and often problematic relationship between lesbianism and the mainstream, however, definite mainstream interventions and behind-the-scenes movements can be seen in recent times. An example can be found in the film *D.E.B.S.* (2004), a low-budget but then mainstreamed teen action-comedy that featured a lesbian couple, was distributed by Sony, and resulted in its lesbian writer/director (Angela Robinson) being offered a deal to direct Disney films.

The perception that the progress of lesbian rights and cultural visibility has suddenly come into fashion, as seen in Alison Glock's pronouncement that since 1999 "lesbianism became hip" (26), is not at all unusual in the discourse circulating in the popular press. What is overlooked by the notoriously short-memoried media critics is that lesbianism "suddenly becoming hip" has happened before. "Lesbian chic" came and went in the 1990s. What can easily be forgotten under the glare of such admittedly very different media spotlights is that lesbian cultures, and lesbian cultural productions, have been present and continue on whether or not the mainstream ascribes them the status of fashionability.

The cultural interventions that visibility politics have made are undoubtedly great. However, some critics still question the nature of visibility as a source of power. For example, Vaid argues that more wide-reaching political change must be made in order to render homosexuals truly equal rather than existing in a state of "virtual equality" (4). Nicki Hastie, in a discussion of lesbian chic on British TV in 1994/1995, makes a point that is still very important a decade later:

> Contradictions between being seen and being heard, and between visibility and power come to the fore when exposing the gaps between characterisations of lesbians and real lesbian lives. . . . The high-gloss popularity associated with "lesbian chic" comes into view only when women are considered to be playing a part and to look the part. . . . As lesbian characters explode onto the TV screen and "lesbian" stories hit the headlines, it becomes imperative to question the gaps between visibility and empowerment. (37–38)

POLITICAL DEBATES

For better or for worse, the primary gay political organizations in the United States decided in the early nineties to focus on normalizing issues such as gays in the military[2] and, later on, gay marriage. Most visible recent gay activism and public debates has been on this front—and this is reflected in a great deal of contemporary cultural productions that feature gays and lesbians. Warner argues that "[s]ince the 1993 March on Washington, marriage has come to dominate the political imagination of the national gay movement in the United States" (84), and with increasing numbers of countries recognizing gay marriage or partnerships and the increasing backlash against it in other parts of the world, the issue will presumably remain the most iconic in the gay and lesbian political movement well into the noughties, together with gay and lesbian parental rights and visibility.

Much of the discourse published about gay marriage and partnerships in English circulates around the particular situation of the United States, but the history of positive same-sex partnership and marriage legislation certainly does not center on the United States. Denmark, which introduced equal age of consent laws in 1976, included sexual orientation in their antidiscrimination laws in 1987 and introduced registered partnerships for same-sex couples with similar rights[3] to marriage in 1989 (Jensen 38–39). In 1995, the Supreme Court of Canada found that discrimination on the basis of sexual orientation was analogous to other forms of discrimination, and various provinces began to make adoption rights available to same-sex couples (as well as individuals) during the late 1990s. In 1997 the European Union adopted the Treaty of Amsterdam, which included a policy of nondiscrimination on the grounds of sexual orientation, to take effect as of May 1999 (ILGA-Europe, *After Amsterdam* 6, 12). The Netherlands became the first country to legalize same-sex marriage in 2001.[4] Belgium followed in 2003.[5] Great Britain introduced registered partnership for same-sex couples in 2004, to take effect in December 2005. Canadian provinces began legalizing same-sex marriage in 2003, and in 2005 Canada approved same-sex marriage on a federal level. Spain's legalization of gay marriage in 2005 was further notable because it contained no exceptions in terms of parenting.[6]

In a backlash against same-sex marriage, other nations (such as Australia in 2004) have "amended the law to specifically exclude same-sex couples from marrying" (T. Fox 10). Despite such amendments, same-sex marriage in Australia has gained a certain cultural authenticity, with such magazines as *Cosmopolitan Bride* featuring a gay or lesbian wedding each month: "Due to the positive response to their lesbian editorial. . . . [they encourage] lesbian brides to approach the magazine" (Johns 8). A situation in which changes are made in cultural perceptions of homosexuality long before law reforms eventuate is not an unusual one in Australia. As historian Graham Willett notes in the 2004 documentary *The Hidden History of Homosexual Australia*, broadcast on SBS Television Australia: "it was really interesting that the early movement—you know Campaign Against Moral Persecution and Gay Liberation—all used to say "look, it will be relatively easy to change the laws, but changing public attitudes will be very hard." In fact the exact opposite was true."

Such shifts can be read as a marker of the success of visibility-based campaigns for cultural change or, conversely, as Vaid suggests, of a situation of "virtual equality":

> The irony of gay and lesbian mainstreaming is that more than fifty years of active effort to challenge homophobia and heterosexism have yielded us not freedom but "virtual equality," which simulates genuine civic equality but cannot transcend the simulation. In this state, gay and lesbian people possess some of the trappings of full equality but are denied all its benefits. We proceed as if we enjoy real freedom, real acceptance, as if we have won lasting changes in the laws and mores of our nation. Some of us even believe that the simulation of equality we have won represents the real thing. But the actual facts and conditions that define gay and lesbian life

demonstrate that we have won "virtual" freedom and "virtual" equal treatment under "virtually" the same laws as straight people. (Vaid 4)

The phenomenon wherein "[a] backlash against gay rights swells at the same instant we witness the widest cultural opening gay people have ever experienced" (1), which Vaid observed in the early nineties, seems even more relevant today as *Queer Eye for the Straight Guy* and *The L Word* are broadcast while constitutions are specifically rewritten to deny legal or state-based challenges for same-sex marriage.

Same-sex marriage has rarely been a unanimous desire for homosexual communities. As Warner comments,

> No one was more surprised by the rise of the gay marriage issue than many veterans of earlier forms of gay activism. To them, marriage seems both less urgent and less agreed upon than such items as HIV and health care, AIDS prevention, the repeal of sodomy laws, antigay violence, job discrimination, immigration, media coverage, military antigay policy, sex inequality, and the saturation of everyday life by heterosexual privilege. (84)

While Sullivan claims that "[m]arriage. . . . is, in fact, the only political and cultural and spiritual institution that can truly liberate us from the shackles of marginalization and pathology" (59), many have objected to the notion on various bases. Feminist lesbian activists who see marriage as a patriarchal institution may have little desire to battle to attain same-sex marriage, while theorists such as Warner see the institution of marriage as a discriminatory one that "sanctifies some couples at the expense of others" (82). In the same vein, activists such as Jenni Millbank "think it is vital that we don't help to create a system where unmarried same sex couples are doubly disadvantaged—by being granted less rights than straight unmarried couples and less rights than married or registered same sex couples" (in Fox 10).[7]

In fact, it was not until the major political backlash against same-sex marriage that it became a fairly unanimous rallying point for the gay and lesbian communities. As Alison Bechdel put it at the 2005 Boston Dyke March,

> if I had charted the progress of this civil rights movement, I wouldn't have picked marriage to be the deciding issue in our attainment of legal and social equality. But now that it is, let's keep fighting for it. But more importantly, once we get it, let's not sit back and lapse into a coma of orthodoxy, but work to undermine the false equation of marriage with citizenship.

No one should have to have a state-approved sex partner to be considered legitimate or to get health insurance. (paraphrasing of speech on Alison Bechdel's Web site *DTWOF: The Blog*)

The shifts in public consciousness around marriage as a symbol of equality rather than a hegemonic institution with the ability to be mocked have had the added effect of making it undesirable to critique gay marriage upon political grounds for fear of appearing to be aligned with right-wing homophobes. Take, for instance, the following example: the cartoon "Teddy Bear's Wedding" by gay cartoonist Robert Triptow

> was a failed project from the late '80s anthology, *Real Girl*, long before gay marriage was a news story. . . . *Teddy Bear's Wedding* was almost pulled from *Juicy Mother* because attitudes shifted due to the same-sex marital events of early 2004. [Triptow] feared that readers would no longer understand that [he] was mocking self-indulgence and artifice. They'd probably think [he] was slamming same-sex marriage itself. [His] solution was to modify the original story a little and draw a sequel with a happier ending. (Triptow in Camper *Juicy Mother* 86)[8]

Triptow's self-censorship is indicative of a hegemonic suppression of attitudes against same-sex marriage within homosexual communities in response to the right-wing backlash.

While the politics of visibility and same-sex marriage take up the vast majority of well-publicized lesbian and gay activism in Western contexts, other ongoing struggles around a variety of other issues continue. Lesbians are activists in struggles surrounding access to reproductive technologies, antiwar coalitions, responses to a variety of homophobic attacks or statements, issues of poverty and third-world debt, antiracism, and many other issues. Groups like the Lesbian Avengers, founded in 1992, still have multiple chapters in the United States and do most of their activism on a grassroots level, promoting lesbian visibility through such activities as street theatre and Dyke Marches. Another U.S.-based performance group, the Radical Cheerleaders, visibly includes but is not solely comprised of queer women).[9] In Australia, groups such as Community Action Against Homophobia, which participates in legal actions as well as protests and kiss-ins, and projects such as Outlink, which aim to "put together a network of young rural lesbian, gay, bisexual and transgendered people with a view to establishing better support services for these young people" (Croome), exist in tandem with the more visible and established groups such as New Mardi Gras and many other project-specific political groupings. In nations where homosexuality is less accepted, activism frequently surrounds the more basic issues of survival, coming out, and challenges to incarceration and violence.[10]

CULTURAL TRENDS

In Figure 2.1, we see Bechdel's key character Mo from *Dykes to Watch Out For* bemoaning the cultural phenomena of lesbian parenting and the increasing number of lesbian women coming out as transgender or transsexual men. Although clearly satirizing Mo's prejudices, this strip does point to a clear trend toward both these developments that are perhaps the largest lesbian cultural trends of the late nineties and early noughties. While Mo's lament that "30 years of radical feminism, and lesbians are embracing Wilma and Fred Flintstone as role models!" is clearly hyperbolic, what does the dominance of these tendencies suggest about contemporary lesbian culture?

Critics such as Warn in "TV's Lesbian Baby Boom" and Judith Halberstam in "The I Word" have written of their dissatisfaction with lesbian story lines that feature weddings and babies. Indeed, with the exception of the lesbian characters on *Buffy the Vampire Slayer*, there are few lesbian televisual representations that do not include characters trying to get pregnant or having children. An interesting example of the dominance of such tropes can be seen in the third segment of the three-part lesbian telemovie *If These Walls Could Talk 2*. After examining the problems associated with the closet in "1961" and the divisions among feminist and lesbian communities in "1972," it presents a lesbian couple attempting to have a child in its "2000" segment—positing this as the defining issue of the times. And this trend is not isolated to the genre of television. The Irish film *Goldfish Memory* (2003), for example, features a lead lesbian character who sleeps with her gay male best friend, finds herself pregnant, and brings up the child with her new (female) partner, while the U.S. documentary *Making Grace* (2005) follows the journey of a lesbian couple from their decision to have a child through the pregnancy and into the child's early years.

While assertions that the endemic scope of these story lines is a way to avoid presenting lesbians as sexual beings may indeed be correct and, I would add, is indicative of a certain normalizing, "just like us" political strategy, the fervor of Warn and Halberstam's detraction of such baby story lines does not take into account the cultural and political currencies of these stories.[11] In accordance with the push for marriage rights and political tendencies toward assimilation, together with the growing efficacy of and access to reproductive technologies, lesbian couples having

Figure 2.1 From "The Misanthrope" (#351, *Dykes and Sundry Other* 37)

children is becoming an increasingly commonplace occurrence. According to the 1990 U.S. census, "[o]ne in five lesbian coupled households included children under the age of 18" (Bradford et al. 28), while the 2000 census found "34% of female unmarried-partner households (i.e., lesbian and bisexual female couples) . . . had at least one child under the age of 18 living with them" (NGLTF 1). From 1999 onward there has been a veritable explosion of how-to "manuals" on pregnancy and parenting for lesbians,[12] together with increasing textual resources for children brought up in gay and lesbian families.

The contemporary period also displays a definite cultural trend toward masculinity within lesbian communities, whether this is performative masculinity (drag kings),[13] female masculinity (bois, butches), or male masculinity (transgender or transsexual men). A key trend within lesbian cultures of the noughties has been the continuing emergence and growth of drag kinging. While once butch-femme cultures were considered by theorists as the camp equivalent of drag queens for gay men, drag kings have recently achieved significant cultural and academic prominence, and the number and popularity of drag kings at lesbian events has steadily risen since the late 1990s. An annual International Drag King Extravaganza has been held since 1999, and it incorporates an academic conference alongside its workshops and performances, bringing "traditional academic scholarship on drag in dialogue with the public intellectual work that [is] taking place on drag king stages across the country (and as we began to soon learn—around the world!)" (Donna Troka 2). Publications such as Halberstam and Del LaGrace Volcano's *The Drag King Book* and *The Drag King Anthology* edited by Troka, Kathleen LeBasco, and Jean (now Bobby) Noble have built up drag kinging as a subject of academic interest, further assisted by the contemporary interest in studying masculinities. In the 2000s, lesbian magazines such as *Lesbians on the Loose* are not shy about putting drag kings on the cover (see the August 2005 issue, for example), and the first issue of the "dyke sex magazine" *Slit* featured "Kings" as its theme (2002). Documentaries such as *Venus Boyz* (2002), which explores both drag kings and transgender men, and *Drag Kings on Tour* (2004) have screened at many queer film festivals and are prominently advertised on lesbian Web sites. Even minifeature films such as *The Undergrad* (2003) have casts comprised almost entirely of drag kings performing as gay and straight men. Each of the texts examined within this monograph engages with lesbian representation or culture and contains drag kings: The U.S. version of *Queer as Folk* displays a drag king show during one of its few lesbian events (2.5), *The L Word* features drag king character Ivan, Lois from *Dykes to Watch Out For* frequently performs as drag king Max Axle, and *Gurlesque* has featured drag king acts on a number of occasions.

Alongside this growth, there has been a steady increase in the cultural prominence of transgender and transsexual individuals within lesbian communities. Lesbian communities have historically been fairly unaccepting of transsexual lesbian women. While growing awareness of this issue during the 1990s has meant that some lesbian communities have become much more supportive of transsexual lesbians, others have continued to resist accepting transsexual women, most prominently through barring them from women-only spaces. Perhaps the biggest skirmish on these border wars in recent years has been the question of whether transsexual women should be allowed to attend the iconic Michigan

Women's Music Festival; and a protest group called "Camp Trans" has since 1994 been set up outside the festival in protest against the "women-born-women" policy.[14] In Australia, organizers of Lesfest 2004 sought and gained an exemption under the Equal Opportunity Act to reserve their festival for "[l]esbians born female," whom they claimed "have become a minority group in the gay and lesbian community, and their culture and identity is in danger of being subsumed and invisibilised" (Weaver 12). The festival, which was challenged by groups such as WOMAN, did not take place due to a technicality of their exemption (Weaver 12–13). The Canadian film *Better than Chocolate* (1999) fictionalized these disputes in the form of the overwhelmingly sympathetic transsexual lesbian character Judy.[15] Judy, a cabaret singer whose performances include an original song entitled "I'm Not a Fucking Drag Queen," is accosted in the bathroom of a local lesbian bar by a hyperbolically presented flannel-clad lesbian (flannel being the *de rigueur* signifier of "old school" feminist lesbians among the post-sex-wars lesbian generation), who shouts at her that she is not a woman.

This issue has become somewhat displaced in recent times by the increasing cultural prominence of lesbians identifying as genderqueer or female-to-male (FTM)/transgender, some of whom continue to identify as lesbians, or lesbians who transition into transgender or transsexual men. Prominent lesbian celebrities who have transitioned in recent times include sex wars activist and erotica writer Patrick (formerly Pat) Califa, who now writes an advice column in the U.S. lesbian magazine *Girlfriends*, and photographer Del La Grace Volcano (formerly Della Grace). Although "transgender" was initially coined as an overarching term to be expressive of and to seek to unite all those whose gender expression is nonnormative, a definite rift is emerging between some groups and individuals who identify as transgender and those who identify as transsexual, in terms of both ideological outlook and political agendas. Mirha Soleil-Ross argues that "[w]e, as a so-called 'movement,' have extreme disagreements on many fundamental issues (examples: Who should be included in men/women's spaces? Should GID [Gender Identity Disorder] be delisted from the DSM [*Diagnostic and Statistical Manual of Mental Disorders*]?)" (in Namaste 97). Ideologically speaking, a major issue under contest is the question of gender binaries. Various members of genderqueer and transgender communities seek to deconstruct gender binaries, while various members of transgender and transsexual communities have no desire to deconstruct binary gender but rather wish to simply assert their positions upon the opposite side thereof.[16]

The representation of transgender and transsexual people within lesbian cultural products is also beginning to rise. *Dykes to Watch Out For*, known for its cultural diversity and its willingness to address issues in contemporary lesbian culture, introduced a story line in 1994 wherein Mo was forced to grapple with her attitudes toward lesbian cultural space via the transsexual lesbian character Jillian ("Au Courant" #193 *Unnatural* 52–53, "Lime Light" #194: 54–55). In 1999, Lois is attracted to transman Jerry ("I.D. fixe?" #323 *Post* 58–59, "flow state" 89–143), and the two later become friends. More recently, Bechdel has introduced a child character, Jonas, soon to become a girl named Janis in the 2004 strip "Siren Thong" (#446 *Invasion of* 102–3). *The D Word*, a spoof of *The L Word* produced by Dyke TV, features the character Dex (played by Geo Wyeth), who is in the process of transitioning, while even *The L Word* includes the transition of new character Moira/Max (played by Daniela Sea) during its third season.

On the other hand, popular mainstream depictions of contemporary lesbian culture center on extremely feminine representations (see, for example, *The L Word*). Does this mean that representational space (albeit in very different spheres) is slowly opening up for both masculine and feminine expressions of lesbianism? In a recent interview with prominent Sydney drag king Sexy Galexy, she commented,

> I'd always thought of myself as a drag queen because I've always loved glamour and dressing up in jewellery and makeup. In a way I was forced to be a drag king because girls can't be drag queens. . . . When I was in the U.S. I had other drag kings ask me where my masculine side was. But I'm just being myself. In drag you can be anyone or anything: it doesn't have to be gender specific. I love being a little raving poof. A lot of the young girls on the scene are little poofs in dyke bodies. We're embracing our femininity again. ("King for a Night" 18)

It is a curious thing when a feminine dyke must access her inner gay man in order to embrace her femininity. The dearth of cultural space for feminine lesbians interested in performance indicated by this statement is a particularly interesting one because it is reflective of a great deal of queer theory that requires gendered "crossing" for access into the realms of that which is deemed 'truly queer.' Perhaps some balance can be found in the current trend for lesbian burlesque, whose performers are predominantly feminine women. According to historian Tara Vaughan Tremmel, "Culturally, [lesbian burlesque is] a counterpoint, in that it gives feminine women a stage. It's a place that complicates the idea of feminine gender and sexuality" (in Zambreno 2003: n.p.). And as lesbian burlesque performer Glita Supernova asserts, "If the drag queens can do it, we can do it better . . . we know what is fucked about with women and society" (in Meredith and Domino 16). Nonetheless, there remains a certain discomfort with the idea and practice of lesbian femininity, and this is reflected by much queer theory as well as theoretical discussions surrounding mainstreaming and "lesbian chic."

Televising Lesbianism

In order to ground my analysis of *Queer as Folk* in Chapter 3 and *The L Word* in Chapter 4, I would like to offer a brief (and certainly not comprehensive) outline of some key contextual texts, trends, and characters in lesbian representations on television. While I will offer some historical examples, I will not give a detailed historical analysis, as this has already been undertaken in other works, most notably in Keith Howes's *Broadcasting It*, Steven Capsuto's *Alternate Channels*, Larry Gross's *Up From Invisibility*, Stephen Tropiano's *The Prime Time Closet*, part two of *All the Rage* by Suzanna Danuta Walters, the introduction and conclusion of Nancy San Martin's dissertation on "Queer TV," and Ron Becker's *Gay TV and Straight America*.[17] It is important to take note that on television the frequent sidelining of the sexual expression of consciously lesbian characters coexists with a voyeuristic oversexualization of "lesbian" kisses between narratively heterosexual women (most famously on *Ally McBeal*). This dual impulse of simultaneously fetishizing the lesbian and erasing visible traces of her lesbianism is certainly not unique to television, but it is a particularly important aspect of lesbian depiction on television. In this section I will focus only on those depictions that have included at least one avowedly lesbian or bisexual character or person, ignoring such sensationalistic televised "lesbian" encounters as Madonna kissing Britney Spears and Christina Aguilera at the 2003 MTV music awards.[18]

While one-episode lesbian and bisexual characters have long appeared on television, various ongoing characters have also been emerging since the late 1970s. These have included the bisexual reporter Vicki Stafford on *The Box* (1974–1977), various lesbian characters in the Australian series *Prisoner: Cell Block H* (1979–1986), Marilyn McGrath as a worker in a women's health clinic in the U.S. series *Heartbeat* (1988–1989),[19] bisexual lawyer C.J. Lamb on *L.A. Law* (1986–1994, character appeared 1990–1992), bisexual Nancy in *Roseanne* (1988–1997, character out from 1991), the "enormously sympathetic" (Wilton 81) lesbian Beth Jordache in the U.K. soap opera *Brookside* (1982–2003, character appeared 1993–1995), Carol and Susan on *Friends* (1994–2004), Rhonda and Suzanne on *Relativity* (1996–1997), Dr. Maggie Doyle and Dr. Kerry Weaver on *E.R.* (1994– , out 1996 and 2000 respectively), Ellen on *Ellen* (1994–1998, out 1997), Nikki Wade and Denny Blood in *Bad Girls* (1999–2006, characters appeared 1999–2001 and 1999–2003 respectively),[20] among several others. Telemovies have also been a major vehicle for addressing lesbian issues in the televisual realm, generally as vehicles for dealing with particular contemporary "issues" such as lesbian parenting (*A Question of Love* 1978, *Other Mothers* 1993) or gays in the military (*Serving in Silence: The Margaret Cammermeyer Story* 1995).

There have also been several television programs (that most histories fail to note) produced by lesbians themselves. The progressive Channel 4 in the United Kingdom, for example, hosted two seasons of a four-night "landmark mix of newly-commissioned programs, feature film premieres, short films and documentaries" entitled *Dyke TV* in 1995 and 1996 ("Tribute to Channel Four").[21] And while the U.S. *Dyke TV* unfortunately stopped broadcasting in 2007, it

> aired for the first time in 1993 on a lone public access cable (PAC) channel in Manhattan. Today, the show is broadcast on PAC stations and the independent satellite channel, Free Speech TV, in 78 cities and 23 states across the US, which have a combined subscriber base of 4 million viewers. Episodes feature current news and events, political commentary, arts, health, sports and other issues of interest to lesbians in the US and beyond. (*Dyke TV* Web site)

Various queer time slots or "news magazines" have also been broadcast, including the Australian multicultural broadcaster SBS's *Out* (1992–1995), which initially drew on material from the Channel 4 series of the same name and then added more Australian and multilingual content (Searle). In recent times, online video streaming has become a popular method for disseminating low-budget community-produced "television" programs, with such groups as Out of the Closet TV enabling the online broadcast of lesbian films and documentaries, footage from protests and comedy nights, and Web-based soap operas such as *The Complex*. The advent of queer cable channels in the United States, Canada, and France has further contributed to lesbian content on television screens, though this content is frequently only available to specific regional and socioeconomic groups.

The initial years of the twenty-first century have been of great significance to the history of lesbian representation on television. The year 2000 alone saw ongoing lesbian characters in new series such as *Queer as Folk* and *Dark Angel*, the coming out of characters in series such as *Buffy the Vampire Slayer* and *All My Children*, and a new lesbian character in *E.R.*, among other ongoing and one-off characters. This represents a significant expansion of lesbian representations on television, both in number and in terms of representational quality. To construct a triumphalist narrative of millennial progress here would, however, be somewhat naïve considering that each of these characterizations is largely marked by the same problems persistently displayed by other lesbian characters on television: the characters are relegated to the sidelines, have insufficient attention paid to their lives *as lesbians*, or have their relationships presented in a desexualized manner.

The sci-fi series *Dark Angel* (2000–2002) featured the character Original Cindy (Valerie Rae Miller), who was certainly an original and contextually unusual character, particularly in light of the unspoken rule that allows lesbian characters to be different but not *too* different. An African American, her lesbianism was presented matter-of-factly from the outset of the series, and she was unambiguously sexual, often leaving scenes in pursuit of various "shorties" (contextually, slang for 'lesbian' in the series) and occasionally straight women, all of whom she appears to be able to seduce. However, one does not see much actual interaction between Cindy and these ladies, and these moments are certainly side dishes to the main character Max's (Jessica Alba) missions of rescue and vengeance and her on-again, off-again (heterosexual) romance. In fact, the only time a relationship of Cindy's is depicted on screen is in "Shorties in Love" (1.16), wherein her ex-lover Diamond (Tangelia Rouse) returns for one episode. Their lustful reunion (mainly off screen) is broken by Max's suspicions of and confrontation with Diamond, who it is revealed has been an unknowing participant in human testing of biological warfare agents (a story line not unusual for the series). Max then assists Diamond in locating the man responsible for her disease. As boils break out on her face and she is clearly near death, Diamond delivers a literal "kiss of death" to the designer of her own death. San Martin also analyses this episode in "Queer TV" (227–32), reading the "kiss" as indicative that in the series "black lesbian sex and desire thrive only outside the (literal) restraints of white masculinity and white power" (232).

Also in 2000, *Buffy the Vampire Slayer*'s (1997–2003) popular geek and novice witch Willow (Alyson Hannigan) met the lesbian witch Tara

(Amber Benson) and, over the course of season four, fell in love with her. Willow and Tara's noticeable lack of on-screen sexual activity or even pronounced attraction has been blamed by the creators on network restrictions and saw some improvement with the season six shift from "family network" Warner Bros to UPN, a cable network. In the seventh and final season, Willow's sexuality was further explored, and she was finally depicted in the first actual on-screen lesbian sex scene on American network television with new girlfriend Kennedy (Iyari Limon)[22] during the final season of the series.[23] *Buffy the Vampire Slayer* proved very popular as a subject of academic attention in general, and several articles and conference papers have explicitly engaged with the lesbian content of the program.[24]

Less critical academic attention has been paid to *Queer as Folk*, though this attention has been expanding in recent years.[25] As the first drama series explicitly dedicated to queer life, *Queer as Folk* (U.K.1999–2000, U.S.2000–2005) broke many barriers in the representation of gay men (and, in its U.S. incarnation, lesbians) on television. While the U.S. remake featured the lesbian characters Lindsay and Melanie in a more developed manner than its U.K. predecessor (perhaps simply the result of having more screen time at their disposal), the characters are still marked by their characterization as asexual or sexually frustrated characters and as foils to the more important narratives of the gay men of the series, as will be further examined in Chapter 3.

In the same year, rival cable networks HBO and Showtime each released a three-part telemovie depicting stories from the lives of lesbians (*If These Walls Could Talk 2*) and a lesbian and several gay men (*Common Ground*) in two earlier decades and then the year 2000. While the earlier two stories presented the impacts of homophobia and the denial of rights to lesbians and gays, the difficulties of coming out, and the conflict between heterosexual feminists and lesbian-feminists and lesbian-feminists and butch-femme communities, the year 2000 stories focus on a lesbian couple attempting to get pregnant and the wedding of two gay men. By posing the year 2000 as a definitive moment of integration, these films present a view of the progress of gay and lesbian history as inherently progressive, with 2000 being the year in which gays are assimilated into the final sphere from which they have been excluded: "the family." It also displays an assimilationist view of gays and lesbians by depicting "family" as the key agenda and thereby presenting a "just like heterosexuals" view of homosexuality. The affirmative narrative and positioning of each of the year 2000 segments also insinuate that the problems depicted in earlier periods/segments (including denial of inheritance rights to same-sex partners where there is no will or a lack of spousal recognition by hospitals; attempted closeting of butch lesbians for political gain; military dismissal; being forced to leave country towns; and the threat of dismissal and subsequent closeting of homosexual schoolteachers) are no longer issues when, in fact, each of these symptoms of homophobia depicted in the earlier narratives are still in effect today.

The 2002 BBC dramatization of Sarah Waters's dramatically successful lesbian novel *Tipping the Velvet* is most unlike the other examples I have mentioned; in fact, it is unlike almost any other example of televisual lesbian representation. Its portrayal of 1890s London—replete with music hall male impersonators, Natalie Barney-esque aristocratic lesbian salons, a working class butch-femme bar, linkage between lesbianism and prostitution, portrayal of grassroots political activism, and an historicization of an active lesbian culture—is undeniably unique. Stylistically, *Tipping the Velvet*'s lavish design and baroque storytelling technique are differently suggestive to the "realist" strategies generally employed in depictions of lesbians on television. Like the BBC's previous adaptation of Jeanette Winterson's lesbian novel *Oranges Are Not the Only Fruit* in 1990—in response to which "[t]he BBC received six million phone calls . . . more than any other serial at the time" ("Get this filth . . . ") and a viewership of "around six million" (Hallam and Marshment "From string of knots . . ." 143),[26]—*Tipping the Velvet* also proved to be an extremely popular miniseries. Waters' Booker Prize–nominated novel *Fingersmith* was also adapted for television by the BBC in 2005. *Fingersmith*, also set in Victorian England and featuring a lesbian relationship, garnered a viewership of 5.1 million for the first episode screening on the BBC ("Touch of Velvet" 28) and has since been exported to ABC television in Australia and BBC America. The BBC is indeed no stranger to lesbian plot lines in its period-drama broadcasts; for example, *Tenko* (1981–1984) included "British drama's first regular lesbian character" in 1981 ("Get this . . .").

Lesbians have also been increasingly visible on recent "reality" TV.[27] In 2002, Fox Chicago broadcasted a reality TV program entitled *Experiment: Gay and Straight* that involved locking five heterosexual and five homosexual people (including two lesbians: one young lesbian and one married lesbian attempting to have a child) in a house together and forcing them to discuss issues surrounding sexual orientation and discrimination. In other, nonspecific depictions, focus on the person's or character's lesbianism is frequently limited (often via editing) or played up as a titillating or menacing factor. The 2004 season of *Survivor* featured not one but two lesbian contestants: Ami (announced as an ex-*Playboy* model) and the older "rancher" Scout, both key forces in an large alliance of women who, for the first time in the series' history, *almost* made it till the end (until Scout broke the alliance). While both women came out early during the series' taping and their homosexuality was widely known among the *Survivor* audience, Ami was not outed on the show until episode eleven.[28] The reasoning behind such editing is perhaps that, as Malinda Lo has suggested, "the show's producers want[ed] viewers to believe that the women's alliance [was] being held together by Ami, who [was] widely perceived to hold an uncanny—and suggestively homosexual—influence over the women" in order to "suggest that she [was] hiding something . . . which feeds into an image of her as manipulative and cunning" ("Woman Power").

Kitten, who appeared on the U.K. *Big Brother* in 2004, was quite different from the carefully neutral, apolitical lesbians usually seen on television. Kitten cites her reasons for participation in the series as "getting a political platform, to be able to discuss issues like war, women's rights, global oppression of minorities and the mistreatment of others less fortunate than ourselves . . . [and] fuck over *Big Brother*." She proceeded to present this platform by employing such tactics as "[d]aubing political slogans on the walls with face paint" and "[e]ncouraging her housemates to stage sit-ins." The most significant opposition to Kitten reputedly came from lesbian viewers, who saw her as "a bad representation of lesbians" (all quotes from interview, Richards 29). In contrast, Ivana, a contestant on the American *Big Brother* in 2005, made a comment reminiscent of Ellen Degeneres's infamous statement that lesbianism "sound[s] like somebody with some kind of disease" (in Handy "Roll Over" 78) with her assertion that "I just say 'I like girls'. . . . I don't say 'I'm a lesbian.' Such a vulgar word. I'd rather someone say 'she's gay'" (quoted in Warn "CBS's *Big Brother* . . .").

Another significant example was contestant Kim on *America's Next Top Model* (Cycle 5, 2005). Openly announcing her sexuality at her audition, her female masculinity was both a contrast to the other contestants and a source of constant struggle. The judges, including genderqueer male judge "Miss Jay," alternated between exhorting her to be more feminine, as seen in Miss Jay's declaration, "you've gotten on the gender-bender bus and now you've got to get off" (5.7), and asking to see Kim's "authentic," masculine self (Tyra Banks 5.4). During her time on the series, Kim also shared kisses and a brief romance with a fellow contestant, the declaratively straight Sarah (at Sarah's instigation). While lesbians on reality series have mostly been restricted to contestants, Bravo's *Queer Eye for the Straight Girl* (2005–)—a spin-off of *Queer Eye for the Straight Guy* (2003–)—includes lesbian "life-consultant" Honey Labrador. *The L Word*, the first high-budget lesbian-focused series, initially appeared in 2004, with subsequent seasons slated to continue until at least 2008. *The L Word* is a unique cultural product, both contextually and in its fascinating textual and extratextual deployment of the discourse of "lesbian chic," as will be discussed in more detail in Chapter 4. While screened on a premium cable channel (Showtime) in the United States and thus seemingly accessible to a limited number of economically privileged viewers in certain geographic areas, it has since been screened in a number of nations and, as can be witnessed on the large number of fan Web sites and message boards that have sprung up since its inception, is accessed by a much wider variety of lesbian and nonlesbian viewers via the internet and DVD box sets.

Lesbians and bisexual women also appear to be moving more firmly into television shows targeted at teenagers or "family" audiences. In Australia, early-evening "family" soap opera *Neighbours* depicted a lesbian teenager (Lana) coming to terms with her sexuality during its 2004 season. Lana (Bridget Neval) kissing her heterosexual best friend Skye (Stephanie McIntosh) provoked "outrage" from the usual quarters—"talkback callers and conservative groups"—which prompted executive producer Rick Pellizzeri to comment, "we've tackled breast cancer, drug abuse, postnatal depression, infertility, stalking, behavioural problems for teenagers, police corruption. . . . We are not condoning or promoting homosexuality. . . . Ultimately, we're telling a story about tolerance" (Nguyen 5). This response should give pause to those who herald the simple, confined visibility that story lines in drama series can give as an unambiguous "breakthrough" with their attitude of tolerating but not condoning homosexuality and considering it within a continuum of breast cancer, police corruption, and drug abuse. Lana's character left the series shortly thereafter, and Skye was safely returned to her boyfriend's arms.

In the recent "lesbian" story line on the popular teen soap *The O.C.*, a bisexual female character (Alex) who had previously been involved with one of the main male characters briefly dated the straight/bi-curious ongoing main character Marissa. When screened in Australia, the series was hyped before the season began as containing "hot girl-on-girl action," and the prurience of the advertising campaign continued to increase throughout the season, despite the fact that within the episodes promoted there is no "action" of any kind, except for one kiss that was, in fact, quite tastefully handled. There were also multiple unproblematized conversations about teenage same-sex dalliances being "a phase" by the parental figures of the series, which "rearticulate the notion that teenaged lesbian sexual experiences can be nothing more than a phase of sexual exploration" (Burgess 223). When Alex brings two friends to a school event to fight Marissa's ex-boyfriend Ryan in a display of machismo, advertisements for the Australian screening of the episode proclaimed that Alex was Marissa's "psychotic lesbian lover." Alex and Marissa break up fairly graciously, with Marissa divulging that "I didn't know I'd have to give up my life for you" and Alex acknowledging, "This" (i.e., with her schoolfellows and her ex-boyfriend) "is where you belong." Afterward, Marissa returns to Ryan on the beach, and their figural reunion is overlaid with a song whose main lyrics posit that "it's real love." Alex thereupon leaves the series in order to go back home to her parents—perhaps signifying a return to the normative heterosexual family.

Other television programs targeted at younger viewers and featuring lesbian story lines in late 2005 included the United Kingdom's *Sugar Rush* (2005–2006), the American *South of Nowhere* (2005–) and Canada's *Degrassi: The Next Generation* (2001–). *South of Nowhere* was fairly unique in that it was screened on a youth-oriented network (The N) and two of its major characters, Spencer and Ashley, are depicted as attracted to women. The portrayal of lesbian sexuality is thus not sidelined as it often is in other series where homosexuality is portrayed only in secondary characters or as "issues of the week." As I have argued previously, however, less physical affection is shown between the lesbian couple than their heterosexual counterparts in the series, similar to the double-standard seen in *Buffy the Vampire Slayer* (Beirne "Introduction" 13). In contrast, *Sugar Rush*, an adaptation of a young adult novel, was much more up-front about the sexual desires and activities of its young protagonists, from its portrayal of Kim's lust for Sugar in the first season to scenes of playful bondage in the second. This can to some degree be accounted for by its time slot (late night) and broadcaster (Channel 4), as well as the generally less restricted nature of U.K. television in terms of presenting sexuality.

Despite such advances, some commentators are wary that the increasing presence of queer characters on U.S. network television may be a temporary trend rather than a true shift, though they appear to be fairly entrenched in cable programming:

> [T]he number of lesbian and gay characters on broadcast-network scripted series TV declined this year [2005] to the lowest level since 1996, when the Gay & Lesbian Alliance Against Defamation began tracking such things. . . . There has been a steady increase in the number of gay individuals

appearing in reality shows on broadcast and cable—and some notable advances in cable dramas. But broadcast TV's retreat has been cause for concern. The mixed report card comes at a time when tolerance and understanding are on the rise in the general public, but the LGBT (lesbian, gay, bisexual and transgender) community has become a primary target of the religious right and politicians who see advantage in exploiting its agenda. . . . The overall presence of LGBT individuals on TV has risen substantially in recent years. This past TV season, 14 lesbian and gay characters appeared occasionally or on an ongoing basis in broadcast primetime series, both scripted and reality. On cable, there were 38 LGBT characters, 26 of them in scripted series. (Haley 25)

The virulence with which religious right groups attack programs that feature homosexual story lines can be seen in the campaign of the Melbourne-based Salt Shakers against Channel 7's broadcasts of *The L Word*, which attempted to get advertisers to withdraw from the series (and in some cases succeeded). Although, whether such campaigns have indeed been as successful as they claim to be, as is discussed by Peter Johnson, Channel 7 did delay the screening of the second season of *The L Word*, only beginning to screen it March 2006 and, even then, at midnight. For nations such as Australia, where cable television is less popular and widespread (many popular U.S. cable programs such as *Sex and the City*, *Queer as Folk*, and *The Sopranos* are broadcast there on network television) and which is historically more open to presenting greater numbers of queer characters in its locally made productions, this circumstance seems rather unusual.[29]

New gay cable networks based in the United States such as Logo and Here! at first seemed to focus the majority of their programming on gay men. Logo's flagship show *Noah's Arc* (2004–), for example, which features four gay male characters of color, only displays lesbian characters in one episode of its so-far two-season run, and even in this case, these characters are included only briefly, with their presence largely included in order to create dramatic tension between one of the male couples of the series. Here!'s soft-porn soap-opera flagship *Dante's Cove* (2005–) makes more of an effort at being inclusive—mostly by featuring a number of topless lesbian and bisexual female characters. Even in *Dante's Cove*, however, where narrative is largely a by-product of rather explicit sexual scenes between the main male characters, a hesitation about depicting similarly graphic lesbian sex is displayed. In the first episode, for example, of the two lesbian sex scenes depicted among a myriad of gay male ones, the first only goes so far as to display a woman kissing the stomach of the other, while the second is likewise coy, cutting away after initial foreplay. This tendency to render lesbian sexuality invisible even in the face of explicit depictions of gay male sexuality is not uncommon, as will be argued in relation to Showtime's *Queer as Folk* in Chapter 3. Logo's animated show *Rick & Steve* likewise focuses on its eponymous gay male characters, and its lesbian characters Dana and Kristen are a secondary feature, with much of their story line revolving around pregnancy, which is also the case in *Queer as Folk*.

The situation appears to be improving, however, with lesbian characters becoming more of a presence in ongoing programming such as *Dante's Cove*, and the channels have also produced some specifically lesbian-focused programming. These have included Here!'s *Lesbian Sex and Sexuality* (2007), a six-part documentary series that featured episodes on such subjects as lesbian pornography and lesbian go-go dancers, *Curl Girls* (2007), a documentary about lesbian and bisexual women surfers that turned into a reality television series on Logo, and Logo's *Exes and Ohs* (2007-), a six-episode scripted lesbian comedy/drama series created by Michelle Paradise. As the second scripted lesbian-centered television series on a nationwide United States channel, *Exes and Ohs* diverged from its forerunner *The L Word* in its focus on more everyday, less glamorous characters, and focus on comedy rather than melodrama. Despite being offered on basic cable rather than the premium Showtime, however, *Exes and Ohs* has not yet gained the same exposure or global fan-response as *The L Word*. This may be due to the less explicit nature of *Exes and Ohs*, or perhaps its presence on a gay-focused channel has led to less media exposure. Should the series continue on for further seasons (as yet unconfirmed), and gain international distribution, it may well take on a cultural significance similar to that of *The L Word*.

As mentioned earlier in this chapter, a clear trend in lesbian televisual depictions is that of placing the lesbian in relation to those traditional prototypes of heterosexual feminine desires—weddings and babies. The accusations that the frequent deployment of such narratives panders to assimilationist interests through normalization and allows the program to avoid showing much real sexuality between their characters certainly have some truth. However, these must be tempered by the realization of the almost inevitably negative narrative consequences of lesbian pregnancy and childrearing on television, including infidelity or even the death of one of the partners. *E.R.*'s story line sees character Kerry Weaver suffer a miscarriage, lose her partner (peripheral character Sandy Lopez, who had recently borne them a child), and become consequently embroiled in a custody battle with Sandy's parents. In *Queer as Folk*, Melanie cheats on Lindsay, partly as a result of being denied custody of their son, which results in the couple's temporary breakup. Several seasons later, while Melanie is pregnant with their second child, Lindsay cheats on Melanie, once again resulting in their breakup. In the first season of *The L Word*, Tina's pregnancy seems to violently increase partner Bette's (Jennifer Beals) dissatisfaction with their relationship, and shortly after Tina (Laurel Holloman) miscarries, Bette begins an affair. The second season presents Tina's second pregnancy as an empowering experience for Tina, though certainly not for Bette, who must struggle with having no legal claim on their child. So while the ubiquity of baby story lines would appear to render a certain positive attitude toward lesbians leading "normalized" lives, the actual contents of the majority of these representations do not seem particularly encouraging to lesbians wishing to become pregnant.

In a "Soapbox" column in *Girlfriends* magazine, well-known theorist Judith "Jack" Halberstam declares that

> [i]n fact, lesbians on TV now are no better off than in the 1970s when *Police Woman* aired an episode in which murderous lesbians were accused of butchering retirement home residents. Hey, lesbian killers loose in L.A. actually sounds like a good premise for a TV show. No sperm banks, no babies, no bed death, no big hair, no makeovers—just lesbians, guns, and non-stop action. Bring it on! (Halberstam 2004: 18)

Halberstam's cynicism forms a stark contrast to the numerous articles, commentaries, and even magazine covers that have announced "TV's Gay Heat Wave" (*Vanity Fair* Dec. 2004) as a mark of inclusion, a watershed in gay visibility, and, hence the formula goes, gay acceptability. Activists who mobilized against the portrayals of lesbians as murderous deviants in the 1970s would no doubt be affronted by Halberstam's expressed preference (undoubtedly at least somewhat tongue-in-cheek) for the earlier depictions. Nonetheless, one can empathize with her desire to see lesbians continue to be presented as deviant outsiders as opposed to the sanitized, conventionalized, commercialized lesbians that currently dominate our television screens.

This distinction is an important one in studies of homosexual representations on television—there is little agreement on what exactly constitutes a positive or a negative image. The significance of this division is heightened by a certain tendency to elide a nuanced analysis in the rush to decide whether a particular image is positive or negative, which, although most marked in nonacademic contexts, can also be seen in theoretical critique on the subject.[30] The side to which one belongs in this "positive or negative" debate usually comes down to politics. An easy example can be seen in the deployment of accusations of "stereotyping" in vastly different ways. The charge of "stereotyping" is an immensely popular one among discussants of televisual representations. *Queer as Folk*, for example, is accused of "stereotyping" in its depiction of gay male life as "all about sex" or praised for resisting "stereotypes" by presenting its characters as sexual beings instead of the sexless friends of straight women. *The L Word*, too, is by turns congratulated upon the femininity of its lesbians and criticized for it. Ultimately, both accusations are to some extent correct. The pervasive cultural perception of lesbians as masculine women exists in tandem with pervasive representations and promotion of lesbian images as expressive of an idealized and excessive femininity.[31] Both quite opposite presentations could roughly be characterized as stereotypes (let us not forget that there are real-life dykes who do in fact conform to each "stereotype," as well as many who combine the two), displaying the limitations of the term and thereby its activation in critique as a definitive marker of "negative" representation.

The extreme polarities that emerge in these debates have much to do with the centrality of visibility to gay and lesbian politics. Rightly or wrongly, television in particular has been widely fetishized as the ultimate conferrer of visibility. Martha Gever notes,

> The appearance of lesbians in such prestigious media as broadcast television and large-circulation news magazines is generally interpreted as indication that we have arrived (after all, that is the plot line common to these reports), but it may also generate scepticism or amusement. . . . Defenders of lesbian visibility may find the attention gratifying, but there are others who recommend closer scrutiny of the interests involved, as well as the images produced. (Gever 2003: 24–25)

The scrutiny of these images is indeed warranted in terms of what these images give to lesbian culture, what they say about it, and what they do to it. While my work is primarily concerned with the former, the latter two aspects are no less important. The manner in which lesbianism is presented to mainstream cultures in various textual examples undoubtedly has a significant influence on attitudes toward homosexuality. And the full impact of the relentless commodification of gay and lesbian culture is yet to be realized.[32] It is, however, essential to not make too easy a correlation between the often conflicting representations seen in a particular televisual text and the understandings a viewer will take from it.

SEX AND PORN: THE WARS AND BEYOND

While the numbers of neoconservative gays wedded to the church and the family may be on the rise, there are certainly still lesbian cultures and lesbian cultural productions that embrace the sex-positive ethos that came to prominence during the 1980s and 1990s. Varying and sometimes contradictory by-products of and engagements with the sex wars can certainly be identified in all the contemporary texts I examine in this monograph, and so some introduction to these issues is necessary to ground my analysis of both the specific contemporary texts I am examining and their politico-cultural import. There is, however, a veritable abundance of materials available that analyzes the debates, cultural products, and significance of the sex wars in its various manifestations, and I am wary of an attempt to recapitulate the wealth of pertinent information, especially in such a constricted manner. The following discussion will therefore be neither a complete history of the sex wars nor even a complete analysis of impacts they have had on contemporary culture but rather a brief elucidation of some of the historical and contemporary arguments that are of central resonance to the texts in question, most particularly the various lesbian pornographic films and the lesbian strip and performance club discussed in Chapter 5 and the discourse that surrounds them.

Most histories of the lesbian sex wars[33] situate its beginning at a conference held at Barnard College in 1982. What has now come to be called "the sex wars" or "the lesbian sex wars" were the intense years of academic papers; launches of lesbian pornographic magazines, videos, and performance nights; and political debate that followed, during which the key subjects of controversy were butch-femme relationships, sadomasochism, pornography, and the use of dildos. While many aspects of this debate are deeply intertwined with the ongoing debates over pornography in feminist communities, the courts, and the media, the *lesbian* sex wars were certainly distinguishable in terms, contexts, and results.

The use of the terminology of "wars" is telling; much of the discourse surrounding these debates is written in terms of a deeply pitched ideological battle. Even by the late 1990s—when antipornography feminism had completely lost its perceived cultural consensus, lesbian pornography had proliferated, and the celebration of butch-femme culture was present in both academic theory and cultural practice—the majority of works discussing the lesbian sex wars couch their arguments against a rather amorphous, indefinable "feminist" creature whose arguments are taken to form a undifferentiated hegemonic force against which it is still necessary to struggle (this can be seen in Terralee Bensinger, Engel, and Mary T. Conway, among many others). While there are certainly those who continue to voice their opposition to the sexual practices and products for which these writers argue, to take their views as dominant, or even widely disseminated, is misrepresentative of the current climate of lesbian communities. While pornography, sadomasochism, butch-femme relationships, and dildos may not be the "hot topics" they once were, elements of each have seeped into the mainstream of lesbian cultures, communities of dedicated practitioners and cultural producers still exist, and many appear to have adopted the libertarian approach of noninterference in the sexual practices of others. However, there are a number of cultural products, such as *The L Word* and the lesbian film *Mango Kiss* (2004), that are indicative of a backlash against practices reclaimed during the sex wars. For lesbians, the contemporary calls by conservative gays to police "deviant" sexualities in the name of homosexual liberation are less pervasive, as many of these calls are authored by gay men who do not appear to be aware that lesbians exist in anything other than devoutly monogamous nuclear-family units. The exception to this lack of obvious calls for restriction occurs during discussion of lesbian representation in the mainstream, wherein the majority of lesbian commentary appears to be disturbed by overly sexualized depictions due to their potential to be pornographically consumed by heterosexual men.

The result of the politically volatile climate of the lesbian sex wars was an outpouring of lesbian pornographic products produced by and for lesbians. Early examples include publications such as *On Our Backs* (1984–1995, 1998–present) and *Bad Attitude* (1984–1987) in America, *Wicked Women* in Australia (1987–1996), *Quim* in the United Kingdom (1989-1995), and *Lezzie Smut* in Canada (1993–c. 1998), among many other smaller-scale, 'zine-style productions. It must be noted that most of these magazines were associated with not only pornography but also another key issue of the sex wars: sadomasochism. Tigress Productions produced the lesbian erotic videos *Erotic in Nature* (1985) and *Hay Fever* (1989),[34] while the editors of *On Our Backs* also launched the production company Fatale Video, which produced and distributed a line of pornographic and sex-educational videos. As Tamsin Wilton notes,

> at the time of writing [1996], there is a veritable explosion of productivity in lesbian erotica. Every country with a lesbian community has, it seems, its lesbian sex magazine. From the well-established *On Our Backs* in the

USA or *Wicked Women* in Australia, to newer publications like *Lezzie Smut* from Canada it seems that there is more lesbian-produced erotica than at any time in history. (171)

It was not only the realms of print and video that saw this explosion, however; lesbian strip shows, sex workshops, and sadomasochistic "play parties" gained in both number and cultural visibility.

Engel, in "Arousing Possibilities: The Cultural Work of Lesbian Pornography," feels that this era of lesbian pornography is now past, seeing the advent of lesbian representation in the mainstream as the death knell for lesbian-produced pornographic production, although "leav[ing] open the possibility that these interventions will find new forms and venues which are now in the process of formation" ("Abstract"):

> When I began this project, I had only an intuitive idea that something in the world of lesbian sex-and-porn wasn't quite the same anymore. What I have come to realize over the course of this writing is that while this project began as a contemporary one, it is now an historical one; the moment that is both the subject and the object of this study has passed, or has at least been immeasurably altered. . . . the consequences of not fully participating in the capitalist project played themselves out in the mid 1990s with the rise of "Lesbian Chic." . . . [T]his commodification happens largely at the hands of the mainstream media, and not the hands of the lesbian cultural producers themselves. . . . The result is a glut of lesbian images that are scrubbed down and made as mainstream as possible. . . . In one media frenzy, assimilation has become the dominant politics of huge segments of the lesbian community. So how does participation in this new commodity market undermine, or at least neutralize, the oppositional politics of lesbian pornographic production? The answer lies in the fact that lesbian pornographic productions "fail" to become commodities, at a moment when North American culture at large is turning lesbians into exactly that. As a result, lesbian pornographic productions seem more and more in the late 1990s to represent not interesting and contradictory forms of cultural critique and community building, but rather poor imitations of their more affluent gay and straight bedfellows. (213–20)

Certainly there do seem indicators that this could be the case. A particularly striking example could be one-time *On Our Backs* editor Heather Findley's switch to editing the more mainstream magazine *Girlfriends*:

> The sexual content of *Girlfriends* is nowhere near as bold. . . . Findlay is somewhat embarrassed by her own magazine's conservatism; it recently discontinued the practice of publishing centerfolds in copies sold on the

newsstand. Only subscribers receive the fold-outs. . . . She says the move away from overt pictorials was necessary for *Girlfriends* to grow beyond its San Francisco roots. Sexually oriented ads were also relegated to the back of the magazine, much as they are in *Playboy* and *Penthouse*. "We wanted to keep some of the independence and verve of *On Our Backs*," she says. "But we also wanted a magazine that a lesbian in Duluth would want to read. *On Our Backs* reflected the sex-radical scene in San Francisco. It was fascinating but unidentifiable for 95 percent of the lesbian market." (Andelman)

These moves to include less sexual content and especially the focus upon target markets rather than sexual or community possibilities would certainly seem to be demonstrative of Engel's fears that the commodification of lesbianism means that lesbian pornography as it once was will no longer exist.

However, despite a definite shift in both the cultural significance and the volume of lesbian pornography produced that Engel locates, there appears to be a renaissance of sex-positive culture flourishing in the early years of the noughties. Though the revamped *On Our Backs* may have gained a more commercialized feel, the more marginal/community productions that Engel prefers still exist in the form of such magazines as the Australian *Slit: Dyke Sex Magazine* (2002–). Lesbian pornographic films such as *Sugar High Glitter City* (2001), *Tick Tock* (2001), *Madam and Eve* (2003), and *The Crash Pad* (2005) are still coming out, and new production companies such as the Australian Biz Films are being set up to produce and distribute lesbian porn in "reaction . . . to mainstream portrayals of lesbian sexuality" (Katrina Fox 13). Pornographic performance culture is also a key feature of contemporary lesbian pornographic production, with events such as *Gurlesque* and lesbian burlesque performances more generally and subculture creative festivals (such as Sydney's SheilaFest 2005, which hosted a workshop on producing lesbian pornographic photography) and sex parties all taking part in the maintenance of lesbian pornographic productions.

The mores and values of the sex-positive culture promoted during the sex wars have inexorably seeped into the mainstream of lesbian cultures. One need only look at any given contemporary lesbian magazine (*Lesbians on the Loose, Diva, Curve, Girlfriends, VelvetPark*) to note the changes that the sex wars have wrought upon the mainstream of lesbian culture: advertisements for dildos are always in abundance; dating classifieds often note whether the poster is butch or femme; some magazines feature fashion spreads (*Diva*), including spreads for leather or rubber fetish-wear (*Lesbians on the Loose*) or semipornographic photo shoots (*VelvetPark*); and articles about sexual practice are very common. Perhaps the most significant example of the sex wars entering mainstreamed LGBT cultures is the production and screening of *Lesbian Sex and Sexuality* by gay and lesbian U.S. cable network Here! This six-episode documentary presents an episode that examines the history of the sex wars and the development of *On Our Backs*, Fatale Video, and S.I.R. Productions (significantly only interviewing and featuring the perspective of those from the sex radical side of the debates). Other episodes feature interviews with lesbian video-pornography directors Shine Louise Houston and Dana Dane and with lesbian go-go dancers and discuss such issues as sexual fantasies and polyamory. Surely one cannot argue that mainstreamed queer culture has displaced lesbian sex radicalism when the products of such radicalism are promoted by a gay cable channel.

While a great deal of the initial discourse surrounding lesbian pornography focused upon whether it had the potential for empowering lesbian sexuality or was an indication of internalized sexism, later work tends to center on lesbian sex-cultural products themselves and their significance. Despite some still-vocal opponents of lesbian pornography (the best known of whom include Andrea Dworkin and Sheila Jeffreys), the tendency within the discourse is that lesbian pornography is slowly becoming legitimated as a subject of textuo-cultural study. This is, of course, due not only to the discourses of the sex wars but also to the academic legitimation of noncanonical textual forms to which cultural studies has contributed. The academic discourse specifically surrounding lesbian pornographic products includes such writers as Jill Dolan (1989), Bensinger (1992), Wilton (1996), Conway (1996 and 1997), Engel (2003), and Heather Butler (2004).

Both Dolan's and Wilton's work are still grounded within the political elements of the sex wars debates. Dolan's article "Desire Cloaked in a Trenchcoat" (1989), notably the only article I am aware of that specifically engages with lesbian striptease,[35] discusses questions of pornographic representation and spectatorship. Dolan argues that

> [l]esbians are appropriating the subject position of the male gaze by beginning to articulate the exchange of desire between women. . . . Rather than gazing through the representational window at their commodification as

women, lesbians are generating and buying their own desire on a different representational economy. Perhaps the lesbian subject can offer a model for women spectators that will appropriate the male gaze. The aim is not to look like men, but to look at all. (64)

Wilton, in her monograph *Finger-Licking Good: The Ins and Outs of Lesbian Sex*, examines the history of the sex wars and engages in close analysis of such topics as lesbian sadomasochism, lesbian visibility, lesbian sex gurus, lesbian pornography, and dildos. Wilton concludes that "[l]esbian sexual desire and pleasure are everywhere silenced, denied, belittled, misrepresented and suppressed by those who would eradicate us," however, "[t]he energetic catalyst of queer activism and of AIDS activism, and the need to resist the efforts made to stamp out lesbian sex, have prompted the development of a lively, inventive and sophisticated lesbian sexual culture" (212).

A central and recurring theme in the discourse surrounding lesbian pornography is a focus upon the potentialities of such products to create or reinvent lesbian communities, as can be seen in both Bensinger's essay "Lesbian Pornography: The Re/Making of (a) Community" and Engel's doctoral dissertation "Arousing Possibilities: The Cultural Work of Lesbian Pornography." To Bensinger, this genesis has occurred via "a lesbian pro-sex recontextualisation of pornographic formal structures [that] disrupts dominant representational practices . . . [and] plays a role in the self-referential re-eroticization of lesbian culture," which has resulted in "the definition of community as something that is constructed through the activity of 'making community' (rather than being merely a static descriptive term of *either* location *or* identity)" (72). For Engel, whose thesis examines "print-based North-American lesbian pornography of the 1980s and early 1990s," lesbian pornographic "texts both represent and generate communities and erotic possibilities which stand in opposition to regulatory frameworks of heterosexuality, privatization and individuality" ("Abstract": n.p.).

Conway's interests lie more clearly in the realms of psychoanalytic discussion of the phallus in lesbian pornography (1996) and issues surrounding spectatorship (1997). Butler, in "What Do You Call a Lesbian with Long Fingers? The Development of Dyke Pornography," charts the evolution of lesbian video pornography and "the various permutations of the butch/femme dyad, the dildo, the concept of authenticity, and the idea of creating through representation a discursive place/space that is coded as a specific lesbian zone" (167). Unlike Conway, who is working through psychoanalytic paradigms, Butler feels that "the very idea of the phallic is displaced when the desire represented is lesbian" (184), perhaps pointing toward a long-overdue shift away from the psychoanalytic approach that has dominated much feminist and lesbian theorizing.

While the flurry of political debate on lesbian pornography and pornographic production had subsided by the noughties, academic discussion thereon started to become more intricate and established. Oppositional lesbian sex cultures and productions are still in existence, and their philosophies have seeped into the mainstream of lesbian culture. The issues of objectification raised by the sex wars strangely concatenate with the issues of commodification so relevant to the contemporary cultural and textual milieu—making them of particular interest to the analysis of mainstream texts.

LESBIAN COMIX AND COMICS[36]

While the subject of lesbianism was occasionally hinted at in various comic strips prior to the 1970s, it was the women's liberation movement that enabled the germination of truly lesbian characters in comics (Robbins 1999). The early seventies saw the emergence of all-women comic book series such as *Tits 'n' Clits*, *Pandora's Box*, and *Wimmen's Comix*. Trina Robbins cites her own contribution to the first *Wimmen's Comix* (1972), "Sandy Comes Out," as "the first comic about a lesbian" (91). Mary Wings responded with *Come Out Comix*, "the first lesbian comic book ever produced" (Robbins 92), in 1973/1974, while Roberta Gregory drew "A Modern Romance" for *Wimmen's Comix* in 1974/1975[37] and featured the character "Bitchy Butch" in her series *Naughty Bits*.

1980 saw Jennifer Camper's comic strip *Cookie Jones, Lesbian Detective* published in *Gay Community News* (Sudell). Bechdel's first *Dykes to Watch Out For* strip appeared in the "feminist monthly newspaper" *Womanews* in 1983 (Bechdel *The Indelible* 27), and her first two collections of *Dykes to Watch Out For* were published in 1986 and 1988. The serial anthology *Gay Comix* (1980–1998)[38] provided a space for the publication of lesbian comics, and this was further expanded through an editorial decision by Andy Mangels in 1991 "to always include an equal number of pages by women and men" (Robbins 102).

The 1990s saw continued growth in both volume and productivity of lesbian graphic-narrative artists and was the decade that truly cemented a lesbian comics subculture. Camper published her first collection, *Rude Girls and Dangerous Women*, in 1994 and followed with the graphic novel *SubGURLZ* in 1999. Bechdel launched six collections of her work throughout the decade and a comics autobiography *The Indelible*

Alison Bechdel: Confessions, Comix, and Miscellaneous Dykes to Watch Out For in 1998. Joan Hilty published a comic about "a lesbian superheroine named Immola . . . who belonged to an all-girl supergroup called the Luna Legion" in *Gay Comix* and the 'zine *OH: A Comic for Her, because It's Time* between 1991 and c. 1996 (Hilty in Gay League 2001: n.p.). Many lesbian comics of the 1990s were self-published and distributed in a 'zine format, [39] and 'zine "trading" and cross-promotion among creators facilitated the growth of a community of lesbian comics artists and fans. This was in large part due to the cultural prevalence of the Riot Grrrl/third-wave-feminist aesthetic of "do-it-yourself" in regard to both political issues and artistic production. Diane DiMassa formed Giant-Ass Publishing in 1991 to distribute her 'zine-format *Hothead Paisan: Homicidal Lesbian Terrorist* comics, the complete set of which were collected into a book by Cleis Press in 1999. Leanne Franson started publishing her *Liliane* comics in "minicomic" format in 1992, as well as in various 'zines and collections, and went on to publish *Assume Nothing* and *Teaching Through Trauma* with Slab-O-Concrete in 1997 and 1999. Paige Braddock published her comic *See Jane* (now *Jane's World*) online beginning in 1998.[40]

In the current climate, cartoonists such as Bechdel, Hilty (*Bitter Girl* 2003–), and Braddock (*Jane's World*—nominated for an Eisner award in 2006) are in syndication, and the major cartoonists release books of their collected work through various presses, generally those known for gay and lesbian publication rather than comic book publication. Others such as Franson (*Liliane, Bi-Dyke*) continue to publish in print (*Don't Be a Crotte*, 2004, and ongoing minicomics from 2004) and post their work on the internet. The internet is fast becoming a primary method of distribution, both for those whose work is also published in print (such as Bechdel, Hilty, and Franson) and as a sole means of publication. Like 'zines in the 1990s, the internet is allowing a new generation of cartoonists to publish their work in a noncommercial or barely commercial manner, which allows for greater multiplicity of comics and less restriction on content and form than traditional print publishing in newspapers or through established comics presses. Lesbian comics in this medium are beginning to become recognized, with Justine Shaw's *Nowhere Girl*, a tale of an alienated young woman struggling to come to terms with her sexual identity, nominated for an Eisner award in 2003.

Of the texts I examine in this book, and arguably among its contemporary cohorts, *Dykes to Watch Out For* is the most overtly political. In terms of both genre and the era in which production began (early 1980s), this is not at all surprising. Viewing lesbian comix generally and those contained within the anthology *Dyke Strippers* (Warren) in particular, political engagement emerges as a—perhaps *the*—central trend. As Anne Thalheimer remarks, "Most lesbian comix are oppositional" (154). This can be partly explained by the feminist focus of that generation of lesbian cartoonists, together with the more general situation of comics as venues for social and political satire.

Academic criticism has mainly focused on *Dykes to Watch Out For* or *Hothead Paisan: Homicidal Lesbian Terrorist*, which are among the most widely distributed and canonical lesbian comic strips. Discussion of lesbian comix has primarily been through informal or semiformal networks (internet message boards or 'zines) or through interviews with the cartoonists, brief articles in lesbian magazines, or on web sites (generally reviews of a collection). In recent years, however, more academic work on lesbian comix has been gradually emerging. Robbins's various historical texts have provided some mention of various lesbian comix and cartoonists, but her work concentrates on female artists more generally, so discussion is fairly brief. Her work does, however, provide an excellent overview of the trends, influences, and development of lesbian comix. Dana Heller, Gabrielle Dean, and Liana Scalettar have written articles on particular aspects of various lesbian comics; Kathleen Martindale wrote a chapter on Bechdel and DiMassa within a broader examination of lesbian writing after the sex wars; and Sara Warner is currently undertaking a project that discusses *Hothead Paisan* in relation to lesbian homicidal tendencies. To date, two theses on lesbian comix have also been completed: Tuula Raikas' thesis "Humour in Alison Bechdel's comics" was completed through the University of Jyväskylä in 1997, and Anne Thalheimer's University of Delaware doctoral thesis, "Terrorists, Bitches, and Dykes: Gender, Violence, and Heteroideology in Late 20th-Century Lesbian Comics," was accepted in 2002.

Undoubtedly the lesbian comic that has been most popular with the critics is DiMassa's *Hothead Paisan*. Heller saw *Hothead Paisan* as "reclaiming a lesbian folklore tradition and transmitting it through a form digestible to a post-Stonewall, post-feminist, postmodern generation better attuned to television, tabloid and mass culture images than to the semiotics of women's folklore" ("Hothead Paisan" 27). Martindale reads *Hothead Paisan* both with and against *Dykes to Watch Out For*, seeing *Hothead Paisan* as more "queerly postmodern" (57) than Bechdel's series, which "seems stuck in the lesbian-feminist political vanguard which currently seems out for the count," (58) while acknowledging that

[i]ronically, DiMassa, the cartoonist probably preferred by younger and queerer readers, employs a "feminist" analysis heavily dependent on concepts like "the patriachy" and misogyny in the mass media, as well as language like "male-identified fem-bot" that were old in the 1970s and are usually thought of as heavy artillery in the ancient and outmoded idiolect known as lesbian-feminism. (70)

Dean interprets DiMassa's work through a psychoanalytic lens, seeing *Hothead Paisan* performing "the role of the stereotypical 'phallicized dyke' as the threat of castration which constructs relations to the phallus and thus gender; yet her personification of this abstract threat reveals the tenuousness of the distinction that castration supposedly imprints on the body" (208). Thalheimer, whose general arguments regarding framing I will discuss more fully in Chapter 6, discusses *Hothead Paisan* primarily in terms of drag as an oppositional strategy and "what contexts we evoke when a lesbian-feminist character who is cast as both parodic and superheroic, whose 'crazy displaced anger' the reader is encouraged to forgive, actually *becomes* a rapist in the name of revenge and how our reading ends up reframing those contexts" (154).

The collection of all *Hothead Paisan* issues into a compendium in 1999 may lead to further academic interest through rendering the texts more easily accessible. The appearance of a musical version of *Hothead Paisan* at the 2004 Michigan Women's Music Festival further mythologizes the cultural significance of DiMassa's work. The feature of *Hothead Paisan* that is of primary interest to my work is the manner in which it straddles the worlds of radical feminism and the 1990s—as Martindale points out, *Hothead Paisan*, despite its clearly radical feminist ideas, has been widely embraced by a queer culture and theory that is reputedly thoroughly disenchanted with feminism. This is demonstrative of the fact that "queer culture," and indeed theory and theorists' attitudes towards cultural products, are far more complicated than fashionable narratives of progression would posit. *Hothead Paisan*, for example, arose in the midst of the 1990s Riot Grrrl scenes, which embraced both early radical feminist ideas and post–sex wars queer culture.

There have also been various lesbian characters in mainstream comics. Although mainstream comic publishing houses have historically been reluctant to introduce gay and lesbian characters, leaving such characters and story lines to "alternative" presses and publications, this situation is slowly beginning to shift. The presence of gay and lesbian creators and editors is often influential on the introduction of gay and lesbian characters to mainstream series: an example is Chris Cooper's authorship of "Marvel's first lesbian lead character, Victoria Montesi" (Mangels 4). This progress has also been enabled by the revised 1989 Comics Code.[41] Whereas the 1954 and 1971 codes had hampered the inclusion of gay and lesbian characters in comic books by proclaiming that "[s]ex perversion or any inference [sic] to same is strictly forbidden" (codes reprinted in Nyberg 168, 174), the 1989 code stated that "[c]haracter portrayals will be carefully crafted and show sensitivity to national, ethnic, religious, sexual, political and socioeconomic orientations" 176).[42] Other lesbian characters introduced in mainstream series include lesbian detective Maggie Sawyer, who was introduced to DC's *Superman* series in 1988 and gained a leading role in a miniseries spin-off *Metropolis S.C.U.* (that received a GLAAD media award in 1996, Gay League "LGBT Comics Timeline"), and *The X-Men*'s Mystique and Destiny, about whom it is said that "[d]espite an unseemly amount of sidestepping around the matter, there is virtually no doubt that Mystique and Destiny were lovers" (Byrd). Renée Montoya, a detective character from DC's *Batman* spin-offs *Gotham Central* and *52*, was outed as a lesbian and portrayed in various relationships with other women (see Palmer). A lesbian Batwoman was later introduced as a love interest for Montoya and was mooted to have her own comic book series released. After much publicity, the publishers cancelled the series prior to publication.

"Dykes in Comicland," an article by Lori Selke in *Curve*, included an excellent potted history of lesbian comics, together with some much-needed publicity for contemporary comic creators such as Franson, Gina Kamentsky (*T-Gina*), Hilty, Erika Moen, and others. Selke also notes "the rising popularity of Asian comics usually referred to under the generic rubric of manga. . . . Although explicitly lesbian-themed manga is still rare, readers note a greater tolerance for the implications of strong female bonding than in other comics and dykes are already drawing and sharing their own works via the web" (75). Along with, and perhaps influenced by, the trend toward yuri-manga (girl-girl manga) Selke locates, are the increasing number of nonpolitical lesbian comics that began to emerge toward the end of the 1990s. Such titles as Elizabeth Watasin's *Charm School* (1997–), the story of lesbian witch Bunny and her romantic triangle with butch lesbian vampire Dean and femme lesbian fairy Fairer Than, steer clear of the political streak so notable among their predecessors. This is not to say that political lesbian comics do not continue to exist; rather, it simply points toward continuing opening up and diversification of the genre.

Unlike Selke's work, various texts that discuss gay and lesbian representation in comics, such as Michelle Helburg's article "Comics Offer Fun, Fully-Developed Lesbian, Bi Characters" or the U.S. version of *Queer as Folk*, frequently describe homosexuality in comics as if it only existed in mainstream circles. Helburg, for example, situates the beginning of "the trend" of lesbian and bisexual characters in comics in long-running indie comic *Love and Rockets*, and she sees it continuing in Brian K. Vaughn's nonprotagonist lesbian characters, Catwoman's lesbian sidekick Holly, and comic book spin-offs of *Buffy the Vampire Slayer* that center around Willow and Tara. The extraordinarily exclusionary nature of Helburg's article is rather baffling, centering as it does upon male-authored sidekick lesbian characters instead of the rich and diverse history of lesbian-created lesbian comics.

Gay comics gain narrative prominence in *Queer as Folk* from the second season onward, when gay studies professor Ben Bruckner asks comic bookstore owner Michael to speak to his class on "gay semiotics" in comic books. Michael and Justin later create a comic book (*Rage*) about a gay superhero modeled on the story line and main protagonist of *Queer as Folk* itself, which, at the end of season four, receives major studio funding for production as a feature film. The motif of the superhero becomes the perfect analogy for the affirmative gay message *Queer as Folk* promulgates: as Michael outlines it, both have fears of their "true identities" being discovered, and yet it is known "that superman will survive and save the world. I believe the same about us. That's what the comics are showing me, that despite everything, we'll survive—and we'll win" (2.6). Called upon to locate work whose "narrative ... graphics, cultural references [and] subtextual points of view" might be regarded as gay, Michael mentions *Alpha Flight* and *X-Force* as well as Destiny and Mystique from *X-Men*. This leaves out a multitude of depictions in which the characters' gay sexuality is much more obvious and significant: the periodical anthologies *Gay Comix* (1980–1998) or *Meatmen* (1986–), the work of Tom of Finland (Touko Laaksonen) or Howard Cruse, Rupert Kinnard's urban superheros the Brown Bomber and Diva Touché Flambé, lesbian avenger Hothead Paisan, or even the mainstream-published male superhero couple Apollo and Midnight in *The Authority* (see Ellis 2000), among many others. By maintaining the focus on mainstream comic books with coded or one-off gay characters, the narrative of *Queer as Folk* is able to falsely claim that the eponymous protagonist of *Rage* is the first "gay superhero"

(2.15) and, it is insinuating, even the first major gay protagonist in a comic book, effectively rewriting the history of gay comics in an otherwise ostensibly "realist" program. The writers' familiarity with gay and lesbian comix is suggested by an allusion to Gregory's character Bitchy Butch made by Debbie in 2.11 as well as its adherence to various conventions of gay *and* lesbian comics, such as impossibly buff and desirable heroes, a blurring between comics and pornography, and the exacting of revenge. The astonishing parallels between the lesbian characterization and narrative of *Queer as Folk* and Bechdel's comic strip *Dykes to Watch Out For* are also suggestive of a certain unacknowledged familiarity with alternative queer comics.[43]

It is important to historicize the appearances of lesbian characters in mainstream comics *within* the history of lesbian comics more broadly and thereby perhaps guard against the major flaws Helburg and *Queer as Folk* promulgate. Both mainstreamed lesbian characters and those continuing and new lesbian characters who appear in lesbian-authored texts and alternative newspapers *can* and *do* simultaneously coexist in the present period—the challenge to those who write about lesbians in comics is to ensure that we can *discuss* both simultaneously and take into consideration the influences and histories they draw from one another.

In the mid 1990s, Universal Press Syndicate proposed to Bechdel that she "develop a gay strip for the mainstream dailies. They didn't want *Dykes to Watch Out For*. They said they wanted something less political" (Bechdel in Keehnen). For many of those to whom mainstream lesbian visibility is the ultimate boon and goal of lesbian activism, such an offer would perhaps have been seen as the endpoint of this narrative of lesbian comics. That Bechdel turned down the offer reveals something about lesbian culture and cultural production that many despairing and celebratory commentators alike often do not take into consideration—that some lesbian cultural producers are still committed to politics and that lesbian cultures are nowhere as monolithic as the trend-proclaimers would have us believe.

CHAPTER 3

Two Babies, a Wedding, and a Man

Queer as Folk and "The Lesbians"

"THE LESBIANS," AS THEY ARE FREQUENTLY REFERRED TO IN *Queer as Folk*, generally take the place traditionally occupied by queer characters in film and television; that is, they are peripheral to the major story lines (except where they are necessary as helpers or foils), have little screen time, have less prominent sex lives than other characters, and are often the subject of ridicule both on and off screen. While these factors in and of themselves would not be notable due to their ubiquity, here they are of particular significance because in *Queer as Folk* it is not heterosexuals who are the main protagonists but rather gay men—thereby displacing yet simultaneously enacting homophobic discourse. However, the lesbian characters on *Queer as Folk* are also of historical and representational significance despite their displacement, because they do receive more screen time, have more on-screen sex, and certainly have more focus on their lives as lesbians than in any previous television series.[1] The first section of this chapter takes up the series' construction of queerness and discussion of assimilationism; the second examines the presentation of lesbian sex on the series; the third focuses on the construction of the queer family on *Queer as Folk*, with particular reference to Lindsay and Brian's relationship; the fourth outlines the series' depiction of marriage; the fifth discusses the question of bisexuality in *Queer as Folk*; and the sixth takes up the subtextual butch-femme codings apparent in Lindsay and Melanie's relationship and the subsequent (re)alignment of femininity with heterosexuality and butchness with abjection. Throughout these sections, issues of gender politics, commodification, and the relationships and attitudes these depictions bear toward preceding politico-cultural texts and movements will continue to arise.

In this chapter I focus on the American version of *Queer as Folk* (2000–2005). Over the course of the series, in true soap-operatic style, we see Melanie and Lindsay have a baby, break up due to Melanie's (lesbian) infidelity, get married, have a threesome, have another baby, once again break up due to Lindsay's (heterosexual) infidelity, get back together again after a bombing, and proceed to move to Canada. My focus on the American series rather than the eponymous British series (1999–2000) is simply due to the greater narrative prominence of the lesbian characters in the later series.[2] Although I am not directly engaged with the original British series in this work, several aspects of the lesbian representation in the American series are drawn from their earlier depictions. Lizzie Thynne, in a brief analysis of the lesbian characters on the original *Queer as Folk* within a larger examination of lesbian characters on British television in the late 1990s (208–12), sees *Queer as Folk* as leaving "the lesbians" as they are: "known half-ironically to the other characters, both straight and gay" on "the margins of the narrative as a kind of unattractive counterpart to the funloving boys" (211). Thynne avers that "Romey [Lindsay's equivalent] signifies the opposite to Stuart's [Brian's equivalent] fecklessness, since she is responsible, politically conscious and maternal—not values which are endorsed by the series' ideology" (211). Lindsay (Thea Gill) and Melanie (Michelle Clunie) are certainly a continuation of this paradigm; in contrast to the male characters, they are responsible and maternal—roles that fall upon women generally in *Queer as Folk* whether homosexual or heterosexual, from Michael's outrageous, fag-hag, working-class mother who acts as universal mother to the gay men of the series (Debbie, played by Sharon Gless) to Justin's (Randy Harrison) more sedate yet still protective and responsible mother (Jennifer, played by Sherry Miller). Furthermore, lesbian characters are primarily depicted in the home (with even their social activities often consisting of parties in their home), whereas the gay male characters appear largely in social spaces (e.g., clubs, work, the diner). The "political consciousness" Thynne identifies in the U.K. *Queer as Folk*'s Romey appears to have all but disappeared in Lindsay,[3] and it is not insignificantly displaced onto Melanie, who acts as the bearer of all lesbian signification that *Queer as Folk* deems negative, namely feminism and female masculinity. In the U.S. version, especially from the second season onward, a very clear dichotomy is set up that poses *Queer as Folk*'s vision of "queerness" against an ever-changing and bizarre conflation of various clearly divergent political practices and aims, all of which are characterized as assimilationist by the narrative. Curiously, the series replays the arguments of the lesbian sex wars, inserting gay men into the roles played by the sex radical side (most prominently *lesbian* S/M practitioners, butch and femme women, and female producers of pornography) and pushing the lesbians of the series into the abnegated position occupied in characterizations of the earlier debate by the lesbian-feminists.

CONSTRUCTING THE QUEER, LOCATING THE ASSIMILATIONIST

The very first words of *Queer as Folk* state, that the "thing you need to know is: it's all about sex. It's true. In fact, they say men think about sex every 28 seconds. Of course, that's straight men. Gay men, it's every nine." This voiceover, directly copied from the British series and here spoken by Michael (Hal Sparks), ostensibly acts as a riposte to the desexualizing practices of both the mainstream media and gay rights advocates and groups and indicates in a deliberately inflammatory and titillating fashion the key theme of the series as a whole. As the focus of the series is indeed on gay men, such a comment is unsurprising and potentially unproblematic. Paul Robinson argues that *Queer as Folk* represents "a cultural watershed" by presenting gays as "not just rights-bearing citizens but desiring bodies" (151, 152), and the erasure of female sexuality can be seen as a necessary—perhaps inevitable—exclusion given the limited representational space given to gay characters on television. However, as I will extrapolate throughout this chapter, the lesbian couple on *Queer as Folk* is consistently used not only as an opportunity for many lesbophobic jokes and remarks but also *as an oppositional force* against which gay male culture, posited by the series as queer culture, is constructed and celebrated. Lesbians are presented as sexless, undesiring creatures;[4] as assimilationists and "funkillers"; as interested only in marriage and babies; as secretly desiring and priotizing men if femme-coded (Lindsay); and as being unfair to men if not (Melanie). The sexual economy is thus vitalized in the show as a purely phallic economy; rigid gender boundaries are enforced and notions of queer families are situated at least initially in terms of the primacy of the father.

The focus on sex and oppositional identity formation in the series can be seen most clearly in the manner in which *Queer as Folk* creates and defines queer identity. Defining the term "queer" is a difficult process, not least because the term itself was born out of a simultaneous resistance to definition and a strategic usage of definition. Kate Monteiro and Sharon Bowers suggest that the definition of queer "which seems to most closely honor the [*Queer as Folk*] producers' use of the word" is that of Annamarie Jagose, who sees queerness as "less an identity than a critique of identity" and, further, "always an identity under construction, a site of permanent becoming" (Jagose qtd. in Monteiro and Bowers). I contend that neither this perception of queer as an identity under construction nor popular conceptions of "queer" as a term of multiplicity that resists static definition are evident in *Queer as Folk*. The term "queer" is used in the title and throughout *Queer as Folk* as a declarative "othering" associated with radical visibility politics, an other to both heterosexuality and the desexualized depictions and definitions of "gay" that have arisen in the popular discourse.[5] Unfortunately, the usage of "queer" in *Queer as Folk* also dismisses the politics of multiplicity and coalition building associated with the usage of the term as a unified replacement for the GLBT string through the evacuation of multiple subject positions or identities from the term. "Queer" is reduced to a signification of gay, white men in *Queer as Folk* not only through absences but also through a vigorous and explicit exclusionary practice. Indeed, the series is quick to attempt to implicitly define "queer," and the definition it undertakes is one that retains the signification of "queer" as indicative of socially transgressive sexual practice while diminishing the other significations associated with the term. *Queer as Folk*'s deployment of the term thus has the most in common with that seen in the lesbian sex wars, where "queers" were posed as the antithesis to lesbian-feminism; and like that previous usage, there is a definite (though reactionary[6]) hierarchical edge.

There are those who would certainly disagree with this assessment. Anne Thalheimer, discussing the evolution of understandings of the term queer, asserts that "[i]n contemporary references, such as in the television program *Queer as Folk*, queerness appears as a universalizing term (where being queer is tantamount to being human) rather than a divisive one" (2).[7] Such an understanding is certainly signifiable from the popular mythology surrounding the title: the producers frequently assert that it comes from an old "Yorkshire expression [that] means there's nothing as strange as people. There's nothing as 'Queer as Folk'" (Cowen and Lipman). This explanation of the title would certainly point toward a universalizing definition of queer, with less focus on the libidinal investments of the term and more on the strangeness or otherness of it. This usage is, however, somewhat misleading because there is little within the text to suggest queerness as a multiply applicable term, with even those who would appear to qualify as queer on identificatory bases being (often actively) excluded on the basis of their gender identities or political beliefs. This takes place not only with the lesbian characters but also with conservative gay characters such as Howard Bellweather.[8] The folk themselves are also not particularly strange—with the exception of their bed-hopping proclivities, their lives are fairly normative. It seems more likely that the "queer" in *Queer as Folk* is derived from this saying only after being loaded by the meanings of its usage in the Queer Nation slogan "Queer as fuck" (Healey 182), wherein the sexual and oppositional elements of queer are emphasized over its strangeness or universalizing qualities—instead of strange as people, it is strange as sex, and this further emphasizes the homosexual connotations of queer, which are deliberately obscured in the earlier, universalizing usage.[9] In *Queer as Folk*, not only is queer inextricably defined by sex as suggested by the Queer Nation usage, but it is defined by certain kinds of (gendered) sex from which various people may be barred.

The exclusionary deployment of queer in the series can be seen most markedly in heroically positioned definitional statements such as the following, which perform queer as not-lesbian and even not-dyke:[10]

> [D]on't get the idea that we are some kind of married couple because we're not. We're not like fucking straight people, we're not like your parents, and we're not a pair of dykes marching down the aisle in matching Vera Wangs.[11] We're queers, and if we're together it's because we wanna be, not because there's locks on our doors. (2.6)

"Queer" is familiarly defined through what it is not—here straight or dyke—and what it entails: relationships of choice rather than restraint. It could be argued that this statement, voiced by Brian (Gale Harold), is not indicative of the series' attitude as a whole, as Brian frequently makes statements that are clearly hyperbolic. However, he is also figured as a reliable narrator of society through his willingness to voice truths that people do not want to hear, and the attitude that Brian here makes explicit is more subtly present throughout the series' depiction of lesbian characters.

As the sexual investments of "queer" have been the key signification retained in the term by the text, the exclusion of women from the sexual economy of *Queer as Folk* operates to further code women as "not queer."

Women, both lesbian and heterosexual, are not figured as sexual beings; rather, they are primarily relegated to the roles of nurturer or caretaker (whether as friend or mother). In this way, *Queer as Folk* emulates the sanitized politics of the antiqueer that it criticizes, simply displacing the restriction to another marginalized figure who has historically been dually represented as frigid or licentious. Men in the series can and do "link sex to love and a marriage-like relationship" (Seidman 133), but they are not naturalized to this position in the same way as women. A strange, almost Victorian attitude to women's sexuality thus emerges in the series. This is first clearly elucidated in Michael's internal monologue during episode 1.3: "I realized how different men and women are, and I don't think it has anything to do with being gay or straight—it's that, the way I see it, women know how to commit to each other, men don't. At least, not the men I know." This scene establishes that the distinctions between the manner in which the sexuality of heterosexual and homosexual characters is portrayed in televisual texts with one or two gay characters are, in *Queer as Folk*, to be displaced onto gender—with lesbians taking up the desexualized role.

Some, such as Robinson in his epilogue to *Queer Wars: The New Gay Right and Its Critics*, which discusses *Queer as Folk*, see little problem in this construction. He acknowledges that "the women are no longer engaged in the dialectic of lust and love that is the leitmotif in the lives of the men [a dialectic Robinson sees as "The show's central theme" (152)]. The show thus supports the widely accepted view that gay men are inclined to be sexual adventurers, while lesbians are nesters." (156)

Yet Robinson argues that "Melanie and Lindsay are [depicted as] fully sexual creatures, and their lovemaking is as graphically portrayed as that of the men (causing a certain queasiness among gay male viewers)" (156). Despite his perception of Melanie and Lindsay as "fully sexual," even Robinson must return to the position that "[s]exually, gays and lesbians are shown to occupy separate universes" (157), an argument he supports with Michael's earlier statement about women being able "to commit to each other," which he falsely attributes to Melanie. It seems that the "queasiness" (156) Robinson attributes to gay male viewers (which presumably includes himself) at the prospect of lesbian sex scenes has utterly overridden his ability to assess the volume or quality of lesbian sex scenes presented in *Queer as Folk*, for Melanie and Lindsay's "lovemaking" is certainly neither as frequently nor "as graphically portrayed" as that of the male characters in the series, to which even the most cursory viewing of *Queer as Folk* will attest. Robinson also fails to problematize the "widely accepted view" that "lesbians are nesters," a view most prominently espoused by members of the "new gay right" that his work critiques, failing to take into account the rich history of lesbians as "sexual adventurers." When he goes on to posit that "[m]arriage in the world of *Queer as Folk* is strictly for lesbians—and, of course, for straights" (157), he once again espouses without question the gay conservative perspective (as does *Queer as Folk* itself, despite the series' obvious disdain for gay conservatives and also despite the fact that the most prominent advocates for same-sex marriage are indeed gay men rather than lesbians).[12]

Lesbian characters are also (at times quite contradictorily) posed as assimilationist. At various points in the series, *Queer as Folk* explicitly positions itself as opposed to assimilation—an attitude that initially appears somewhat odd considering that the presence of a television series about gay people is often seen as a measure of integration. The term "assimilation," like the term "queer," takes on a very specific dimension of its generally understood meanings in *Queer as Folk*: assimilationism is seen as attitudes that would closet the sexual aspects of sexuality in the name of integration. Other factors usually seen as assimilationist, such as the commodification of homosexuality or the willingness to closet diversity, are not understood as meaningfully problematic by the series, and in fact, these aspects of assimilation are often celebrated as positive queer political strategies.

The rhetorical conflict between assimilationism and queerness in the series is presented as a war of images and capital between those who try to integrate homosexuals into the mainstream (by "normalizing" and desexualizing homosexual practices) and those who intervene in, sexualize, and "queer" the mainstream. Despite many of the characters in *Queer as Folk* voicing pro-assimilationist views and the text's narrative impulse leaning to some extent toward the "responsibility" and marriage-like relationships favored by assimilationist politics, *Queer as Folk* positions itself as antiassimilationist (at least as far as sex is concerned) throughout the series. In an article quoting producer Ron Cowen listing some of the key issues of *Queer as Folk*, Cowen particularly focuses on "the conflict in the community between the assimilationists and those who want to continue a queer lifestyle, whom Brian represents. I think there's a huge conflict between those two elements right now" (qtd. in Kaiser).[13] This theme has been very clearly observable throughout the series, particularly from the second season onward.[14]

An interesting example of the series depicting Melanie and Lindsay as assimilationist can be seen in episode 3.6. Emmett (Peter Paige) and Ted (Scott Lowell) are planning to decorate their first home in a decidedly colorful and flamboyant style. Melanie and Lindsay advise them to decorate in a more subdued manner in order to fit into their new suburban neighborhood. After some rather monotonous jokes about lesbian style deficiencies,[15] Emmett makes an attempt at a more normative style in dress and decorating, admitting that "maybe Mel and Linds were right; it was too—fruity." Ted responds, "so instead we get assimilationist beige, blend-right-in brown, and make-no-waves grey." Appealing to Emmett's "self respect" and gay pride, Ted invokes Stonewall to convince Emmett to return to his usual campy self: "Look, no matter what happens in there tonight just remember this is what Pride is all about, why our forefathers and fore Drag Queens stood their ground at Stonewall, so that we could buy a house in a neighborhood like this."

It is somewhat curious that a character such as Melanie, who generally does appear to "have pride," who is quite strident about her lesbianism, and who is actively involved in the gay community—defending Vic and Ted against obscenity charges, challenging gay parenting laws, and raising money for the Gay and Lesbian Center or local AIDS hospice—is portrayed as assimilationist in contrast to the Republican Ted, who, with the exception of his predilection for porn, is usually portrayed as the most normative and conservative of the characters, as seen in his worship of the fictional gay conservative Howard Bellweather. That lesbians are denied their central role in Ted's revisionist invocation of the Stonewall riots further excludes lesbians from being associable with gay pride; queerness is presented as a purely male domain, despite the realities of this particular historical situation and lesbians' role in promoting queer rights and pride even within the story line of *Queer as Folk*. Whereas Ted only locates gay men and drag queens as standing "their ground at Stonewall," multiple witnesses of the actual Stonewall riot agree that it was the "heroic fight" of a butch lesbian "with the police that ignited the riot" and that lesbians more generally were the earliest resisters of police repression at Stonewall (Carter 151).[16]

The ferocity with which *Queer as Folk* "takes on" assimilation is particularly interesting in light of the main forms of political activism that *Queer as Folk* promotes—all of which are driven, in one way or another, by money. Visibility is achieved both in and through the text by selling gayness—both as a means of marketing to the straight population through the "sexiness quotient" and as a lucrative consumer market that businesses can tap into.[17] Although such moves do not remove the sex from queerness in the name of assimilation—indeed they revel in it—they do, however, engage in another form of assimilation that *Queer as Folk* fails to take into account. As Sally Munt discusses in her essay on the British *Queer as Folk*, "Gayness has been formulaically rebranded as attractive and aspirational, it has acquired cultural and symbolic capital, it has, through commodification, become *respectable*" (539). This is further complicated by the position of *Queer as Folk* itself as a creation of which Brian himself would be proud—both as a product sold through in-your-face gay sex and as a means for a (heterosexual) cable channel to target a niche and potentially lucrative market, "capitaliz[ing] on gay identity as a desirable (no longer shamed) commodity" (Munt 533). *Queer as Folk* neglects to portray the problematic elements of commodification, such as co-optation, exploitation, and potential misrepresentation, viewing it principally as a positive development.[18] Such accounts also fail to acknowledge that one can only engage in this kind of "market activism" because of the work of many generations of GLBTQI activists who have provided and continue to provide the psychological and legal basis; the support structures; and the groups, festivals, and publications that have helped shape queer communities that can now be both sold and sold to.

LESBIANS, DESIRE, AND SEX (WARS)

Lesbian sex is generally presented in *Queer as Folk* as being located solely within the confines of a committed monogamous relationship. It is also portrayed as being somehow less "queer" than gay male sex: in *Queer as Folk*, lesbians and women, unlike men, in general are not figured as subjects consumed and driven by sexual desire. Melanie and Lindsay are rarely involved in sexual activity in the series, and when they are, the manner in which these scenes unfold is temporally and representationally limited. Perhaps the most marked feature thereof is the constant interruptions they encounter—either within the narrative by phone calls and doorbells or outside the narrative by cutting away to another scene during or immediately after initial foreplay. As a result, we only very rarely see a woman at the height of pleasure, and only once do we see a woman achieve orgasm, once again rendering lesbian sex as mysterious and women's desires as unknowable and undepictable. This cannot be due to coyness or network restrictions, as not only are the male characters frequently depicted during orgasm, but orgasm is situated as central to many of the sex acts—whether by narratively focusing on them (e.g., Justin's squirting scene in the first episode) or simply by presenting sex whose ultimate goal is orgasm. In fact, a "completed act of lesbian sex (with one of the women achieving orgasm on screen) does not occur until Episode 13's one-night stand, thus contextualising the act of female orgasm as an outlaw one" (Monteiro and Bowers). The deprioritizing of orgasm in depictions of lesbian sex is not a new one. Even in a pornographic film made by and for lesbians discussed by Heather Butler, "[n]one of the scenes end with orgasm; in fact, orgasm does not seem to be a preoccupation in this film" (187). When placed in direct contrast to the male, goal-oriented sexuality in *Queer as Folk*, however, the presentation of only foreplay or frustrated sexual acts between the lesbian characters reinforces socially preconceived notions of lesbian sex *as* foreplay, inherently incomplete and lacking.

Unlike Ted's *physical* problem of being unable to have sex in episode 2.9 (caused by an overload of sexual imagery), in Melanie and Lindsay's case, their lack of sex is rather induced by a lack of desire, especially on the part of Lindsay. While these periods cause Melanie a significant amount of frustration and upset, as she discusses with Ted in the first season and expresses to Lindsay in the second, they do not appear to be an issue to Lindsay. In fact, Lindsay frequently yawns or otherwise expresses her lack of interest when Melanie does attempt to kiss or seduce her. One of the few times we see Lindsay initiating what appears to be a seduction, it soon becomes apparent that she is doing so to manipulate Melanie to "give Brian a chance" (1.5)—portraying feminine sexuality as a tool instead of an end in itself.

Nevertheless, *Queer as Folk* certainly does demonstrate progress in depictions of lesbian sexuality on television. As is examined in Chapter 2, earlier portrayals generally presented either completely desexualized "good" lesbians or hypersexualized, usually bisexual, "bad girls." As can be expected from a show not only "all about sex" (1.1) but also specifically targeted at a homosexual (albeit gay male) audience, Melanie and Lindsay do at least take a step forward from the comforting arm pats or chaste kisses seen in *Heartbeat* or most of *Buffy the Vampire Slayer* or the tame titillation served up by sweeps week "lesbian" kisses.[19] At the beginning of the second season of *Queer as Folk*, the producers even appeared to be making an attempt to provide some degree of corrective to the lesbian sexual characterization of the first season, which was soundly critiqued by lesbian viewers. This is engendered both by showing the couple engaged in more sexual encounters and allowing them to respond to various comments made by other characters:

> *Michael:* Because sex is different for men than it is for women, the need is more immediate, more intense. At least that's what I've read.
> *Lindsay:* Where? In *Field and Stream?*
> *Melanie:* Now, just for your information, Lindsay and I fuck like crazy, we pant and drool like a couple of bitches in heat. Our pussies soak the sheets.
> *Lindsay:* And we go on a lot longer than the ten minute tumble you guys call sex.
> *Melanie:* You don't wanna know how many times we get off in a night.
> *Michael:* You're right, I don't. Mom!! (2.8)

This interaction offsets the frequent comments made by Michael and the others as to "the lesbians" perceived lack of sexual desire. It is perhaps further intended to imply that it is the characters themselves, rather than the writers of the series, who hold such views.

The previous conversation is referenced later in the season by Michael, who replies to Ted's exhortation to "end the sex talk, lesbian approaching," with "Are you kidding? Mel and Linds are a couple of sex machines, they soak the sheets!" Here, however, this assertion is undermined with the presentation of a grumpy Melanie and a discussion that quickly turns to the subject of "lesbian bed death" (2.17). While Melanie contends that such a thing "only exists in the minds of cunty gay men," she later has a discussion with Lindsay and Leda in which she says "there must be something to it," to which Leda reassures them that "your battery isn't dead, it just needs recharging." Despite this latter remark and Melanie and Lindsay's later sexual reunification, this story line tellingly reveals the remarkable similarity between the narrative arcs of the first two seasons for these characters. In the beginning of each of these seasons, the couple is shown happy and engaging in a traditional rite of passage (the birth of Gus in the first season and their wedding in the second), after which a period of "lesbian bed death" ensues, as do a variety of other conflicts; an act of extramarital sex occurs (Melanie's affair, the threesome with Leda); the couple is then reunified by someone else (Brian in season one and Leda in season two). It is interesting that within a couple of episodes of their wedding, however, the more sexual characterization of the two, presumably sparked by critique of the sexlessness of Melanie and Lindsay in the first season, suddenly and inexplicably shifts to the couple considering whether they have fallen prey to "lesbian bed death" (2.17). The inference seems to be that gestures at societal assimilation (e.g., babies, weddings) hamper queer sex and thereby, within the schema of the series, queerness itself. During season three, no sex occurs between the two characters, aside from a rather lackluster and rushed insemination, the depiction of which would indicate that it should not be characterized as sex. Season four sees Melanie pregnant and amorous but gaining little or no response from Lindsay, whose interest lies with Sam, while season 5 shows the couple estranged up until episode 5.9, where they share a passionately violent sexual reunion, perhaps inspired by a similar scene between Bette and Tina in *The L Word* (1.13).

Prior to the very clear focus on assimilationism in season five, episode 2.3 was the episode that most explicitly discussed and ridiculed the forces of assimilationism in U.S. queer culture. During this episode, a scene occurs wherein the cochairs of the Gay and Lesbian Center interrupt Melanie and Lindsay having sex, and this is intercut with scenes of Brian penetrating a handcuffed "police officer" with his own nightstick. But while Brian remains sexually uninterrupted by the demands of others (in the form of the persistently ringing phone) as he puts pleasure first, the women not only are stopped but feel shamed of the fact that they were having sex at all and particularly about their use of a vibrator. Melanie and Lindsay's fear that the co-chairs will discover that they were having sex is compounded by their humiliation when the co-chairs discover the vibrator. This is despite the fact that now, over twenty years after the sex wars began, the use of vibrators is fairly widely acceptable within the lesbian community.[20]

Although the scene is presumably primarily included for its comedic value, it is making a clear point about the sex wars and lesbian culture. This can be seen in the fact that the two co-chairs' revulsion is particularly portrayed through Janice, whose generalized horror at sex, and whose attempt later in the episode to disparage Melanie and Lindsay by calling them "femmes," is intended to implicate her as an "antisex," pre-sex-wars lesbian. Janice's positioning as a representative of "an old, bad, exclusive, policing lesbian feminism" (Walters "From Here to Queer" 848) is further visually reinforced by her presentation as less feminine (less makeup, almost masculine pants, and unstyled hair) and less "attractive" (according to conventional ideals of beauty) than the other main female characters, once again setting up the contemporarily popular narrative dichotomy that pits the lesbians seen in lesbian chic—stylish, young, visually indistinguishable from heterosexual women, nonpolitical, sex-positive and "good"—against the dowdy, older, (somewhat) visible, political, feminist, "bad" lesbians.[21] In this representation, through presenting Janice as a supporter of Howard Bellweather and later as a supporter of the homophobic Stockwell (3.10), the two quite clearly counterposed forces of lesbian-feminism and assimilationist gay politics are conflated into a single malevolent force, here quite literally the character of Janice. This perhaps acts as a comment on the frequently cited similarities in perspectives between conservatives and second-wave feminists when it comes to pornography. Interestingly and ironically, the use of style and mainstream sexiness to denigrate Janice in this depiction and as a means of cementing superior status is similar to the manner in which moralism in assimilationist discourse is seen in this episode to create a distinction between insiders and outsiders.

A further engagement with the issues of the sex wars can be seen in 2.6's engagement with attitudes surrounding pornography. Lindsay is horrified by Ted's new occupation as the owner of a porn site. Her pronouncement that "I don't understand people spreading their legs for the whole world to see. I know it exists, but the idea of anyone I know being part of it really bothers me" prompts Melanie to reveal a copy of *Oui*, the porn magazine that she once posed for, explaining that she did it to pay for her college tuition after her parents briefly cut her off when she came out. The camera cuts away from Lindsay's horrified reaction after a series of "oh my god"s, and the next time we see her in the episode, she is lying in bed flipping listlessly through *Sophisticated Bride*. She then furtively opens *Oui* and begins to masturbate. Melanie, who has been watching her unseen, suddenly says, "I thought you didn't like porn," to which Lindsay responds, "I don't, and I like even less the idea of thinking about all those strange men, and probably even a few women, looking at you. . . . I don't like us having any secrets." Despite these protestations, the two proceed to engage in foreplay until the camera cuts to the next scene. This reaction undermines the antiporn perspective in much the same way as Bellweather's assimilationist perspective is undermined—by outing its advocate as secretly gaining pleasure from the very thing he or she criticizes. Interestingly, even though it has been suggested that Lindsay was an activist (presumably a radical feminist) in college, her objection is not framed in terms of a feminist critique of objectification or of the sex industry itself, but rather first as a prudish horror at Ted and then as an assertion of ownership over Melanie's body and past, diminishing both the efficacy of her argument and the legitimacy of her perspective.

There are, however, characters and incidents within *Queer as Folk* that, to differing extents, counter the perception of women as relationship-orientated or assimilationist. Through Marianne in season one and Leda in season two, the text offers the suggestion that women too have sexual needs that must be met, and if they are not met within their relationship unit, then women may look outside it. During the first of these instances—the sex scene between Marianne and Melanie—various narrative devices and camera techniques are used to undermine this encounter as one of sexual need and desire and instead to emphasize its transgressive nature and Melanie's guilt by frequently focusing on Melanie's wedding band and intercutting the scene with shots of Melanie remorsefully crawling into bed with Lindsay.

The threesome with Leda, in contrast, is portrayed as sexy and an inspiration to the couple, yet they later deny that they have both desired Leda for some time, and they attempt to get rid of her as soon as possible. This sexual encounter is further used as a plot device to *restore* Melanie and Lindsay to happy couple-dom. Leda is introduced in the fourth episode of the second season and is quickly encoded as the most markedly sexual and "queer" lesbian character in *Queer as Folk*, going some way toward countering Melanie and Lindsay's representation as the (only) lesbians in the series. Leda seems to be an archetypal representative for the sex-positive lesbian: her relationships consist of one-night stands and she constantly refers to strap-ons, is dismissive of gay marriage, rides a motorbike; and organizes a stagette party for Melanie and Lindsay, complete with lesbian strippers. She undermines homo/hetero boundaries and manages to surprise even Brian with her overt sexuality when she asks him if he has "ever been fucked by a dyke with a dildo," thereby placing the quintessential gay male top into the recipient position (2.7). Later in this episode, Leda is seen dancing suggestively with both Brian and Justin, undermining articulations of both sexuality and gender established by the series. The manner in which this overt sexuality or empowerment takes place is an interesting one, as it appears that in *Queer as Folk* queer positionality is only available to women via entry into the phallic economy and via identification with gay males over and above lesbians or straight women. Leda's frequent and vocal dissociation from and denigration of "the ladies who munch" (also 2.7) strangely shows her queerness as also constructed via opposition to "the lesbians" and, even more curiously, cunnilingus. It seems that the only way the text is able to picture lesbians as queer or overtly sexual beings is to figure them as phallic women or "dildo dykes." This tendency can also be seen in the work of writers such as Michael Warner, on whose list of "sexual outlaws" the only female outlaws consist of "dildo dykes" (32).

The playing of Bikini Kill's Riot Grrrl anthem "Rebel Girl" (1994) during the the stagette party scene further associates Leda with a certain 1990s queer world. As in the mainstream appropriation of Riot Grrrl discussed by Kristen Schilt, *Queer as Folk* through Leda claims a connection to Riot Grrrls while being devoid of the complex feminist politics of Riot Grrrls, which are simplified into straightforward sexual libertarianism. The song is still, however, certainly evocative of Leda's difference from the more normative, somewhat older characters, Melanie and Lindsay. Despite these factors that code her queerness, Leda is still presented within certain parameters of what is deemed to be acceptable lesbian representation by *Queer as Folk*. She may be overtly sexual and a biker, but she is still thin, femme, and conventionally gorgeous—rendering her physical presence much less threatening to the gay men around her than the unnamed butch lesbian in episode 1.3 to whom Michael reacts with fear and disgust. There is also a degree of textual retrieval of Leda back to the realms of monogamous desires. After her threesome with Lindsay and Melanie, she immediately latches onto them and expresses a desire to "settle down" before riding off out of *Queer as Folk*.

In spite of such shifts as introducing Leda or portraying Melanie and Lindsay in a more sexual light during the second season, lesbians are associated throughout the series with the home, marriage, and children—visions of the good sexual citizen—while the male characters' lives are very much that of the bad sexual citizens.[22] This sense of the lesbian as the figure (though not necessarily paragon) of monogamy in the series leads to the major story line revolving around Melanie and Lindsay in season two: their wedding. This event and the long lead up to it enables the series to engage with the current debates both within and outside of the gay and lesbian movements with regards to marriage, voicing different perspectives thereon through different characters.

LINDSAY, BRIAN, AND QUEER FAMILIES

A key source of friction in Lindsay and Melanie's relationship is Brian or, more specifically, Lindsay's relationship with and feelings toward Brian, which do not appear to be entirely platonic. This can be seen from the very first episode, in which Lindsay and Brian share an intimate scene just after Gus's birth (for whom Brian had been the sperm donor). After sending Melanie out of the room to fetch ice, Lindsay and Brian lie together on the single hospital bed holding Gus, framed together as an archetypal nuclear family unit. Lindsay says, "Who would have thought? You and me. Parents," and the following discussion follows:

> *Brian:* I would have fucked you, you know. If I wasn't afraid your lover'd beat the shit out of me.
> *Lindsay:* Stop.
> *Brian:* I mean it. She could take out Oscar de la Renta.
> *Lindsay:* You mean La Hoya.
> *Brian:* Whatever.
> *Lindsay:* Well, you had plenty of chances.
> *Brian:* And I took advantage of a few, if I recall.

When Melanie reenters the room, the two are kissing one another tenderly on the lips. Mel's dislocation from the frame, and thereby the family unit, is indicative of later developments in the season in which Brian refuses to relinquish his parental rights to Gus. It also, however, establishes Lindsay's bond with Brian and sets the scene both for her subsequent preferential treatment of him and for the view of Brian as the "real" second parent of Gus.

The example of Gus' birth is certainly not the only time when Lindsay easily slips into the pretense of being Brian's wife. In the first season, when shopping for a new car for Brian, the salesman presumes that they are a heterosexual couple, suggesting that Brian should not purchase the jeep he is interested in because "fags drive it" (1.8). Lindsay is quite satisfied to go along with this masquerade and indeed appears to relish being mistaken as Brian's wife. Brian, in contrast, is not satisfied to pass, a point he makes by taking the car for a test drive and smashing it through the display window of the car dealership. The most explicit example of Lindsay's interest in Brian occurs in the second season, just after she has married Melanie and they are attempting to get Gus into a preschool. Having been rejected by one school, presumably due to homophobia, Lindsay asks Brian to present himself as Gus' father at the interview", justifying the clandestine nature of her actions to him with the following:

> You've always said it's not a lie if they make you lie. You want Gus to have the same advantages as a kid who's got a mommy and a daddy don't you? You want him to go to the best schools? Get the best education? Then you've gotta help us make sure, despite what other people might think of us, he's not the one who suffers. (2.13)

The discourse of closeting "for the sake of others" is one frequently espoused by Lindsay; the other notable occasion of this is at her sister's wedding. Melanie is somewhat unhappy at being excluded but goes along with it for the sake of her son. Lindsay, on the contrary, gets quite caught up in the whole simulation, using it as an opportunity to determine if Brian reciprocates her feelings:

> Lindsay: (*Looking at him dreamily*) In fact there was even a time, when we first met, when I thought this could have been the reality. Did you ever feel that way?
> Brian: No.
> (*Lindsay looks both hurt and disappointed and looks away sadly.*)
> Brian: Well, you wanted me to be serious, I . . . it doesn't mean I don't love you.
> (*He puts his arm around her then kisses her. As they pull back his eyes are open looking at her, hers are closed savouring the kiss, until their names are called by the principal.*) (2.13)

Their charade is undermined, as is the necessity of such pretense, when Gus does not get into the second preschool, as preference has been given to a child with same-sex parents to ensure diversity. Lindsay then declares, "first thing tomorrow we're gonna look for another school, and this time Gus' parents, Lindsay and *Melanie*, will go to the interview." The fervor of this announcement seems to be in some degree a response not only to the unexpected decision but also to Brian's rejection of her.

Suzanna Danuta Walters' monograph *All the Rage*, which focuses on gay visibility in America, includes a very brief analysis of (seemingly only the first half of) the first season of *Queer as Folk*. Walters echoes the critics' perceptions of the series as groundbreaking: "Not only are all the main characters gay (and straight characters relegated to the outsider role), but *Queer as Folk* breaks through that other stubborn double standard by displaying active and explicit gay sexuality" (Walters, *All the Rage* 121). This praise is tempered by two primary critiques: that "*Queer as Folk* seems to substitute sexuality for community and to imply that gay sexual expression means an absolute erasure of everything else" (122) and that "[t]he lesbians are there—like women in so many films and TV shows—to serve as nutritional supplements to the main course of male (this time gay male) sexuality and life" (124). For Walters, it seems that these two quite disparate critiques ultimately rest on the same issue: her dissatisfaction with *Queer as Folk*'s depiction of family.

Walters, whose ideal vision of family involves "[r]eplacing those illusory and archaic bonds of biology with the hard-working and complicated bungee cords of choice" (232), is disturbed by her perception that *Queer as Folk*'s visions of "gay community and gay friendships—which have provided such creative and vivid alternatives to the standard familial couplings of American heterosexuals—are here made newly invisible" (122) and sees the series' "potential for a compelling depiction of alternative gay families" as being "lost in the nails on the chalkboard rendering of the couple who, curiously, decide to have a baby with Brian—a thoroughly rancid person who just happens to be despised by a member of the couple" (123). While I must certainly agree with Walters' assessment of the frustrating depiction of the lesbian couple, I feel that Walters' narrow views on what may and what may not constitute an "alternative family" are limiting and fail to take into account the multiple configurations of "alternative" queer families. I also disagree with her assessment that "creative and vivid alternatives to the standard familial couplings" are invisible in *Queer as Folk*.

One can, however, certainly understand how such an assessment can be reached, as the assertion of biological primacy in understandings of Gus' parenting is clear in early episodes of *Queer as Folk*. The exclusion of Melanie not only occurs through institutions (though this can be seen when Melanie is excluded from visiting her son in the hospital while Brian is immediately let in) (1.9) but also is linguistically emphasized through the constant reference to *the* father, which acts to exclude Melanie from her role as coparent. Despite this, however, the nuclear family is not the primary unit of family in the series. Biological families are, of course, presented on *Queer as Folk*, but these families are not, however, in any sense traditional; rather, the series speaks to the multiple possible configurations of brothers, sisters, mothers, friends, and ex-lovers that can create family. Within *Queer as Folk* one can see a progressing emphasis on the queer family in its manifestation as an intergenerational group linked by biology, sexual relationships, *and* friendship.[23] Although there are a number of other examples, perhaps the clearest undermining of biological primacy in conceptions of parenthood occurs during Melanie, Lindsay, and Michael's custody battle for Jenny Rebecca at the beginning of season 5, wherein Lindsay, the nonbiological parent, is portrayed as the best parent via her putting Jenny Rebecca's needs above her own by "withdrawing from this custodial circus" and challenging both Melanie and Michael to "spend as much time thinking about [Jenny Rebecca's] needs as you do about yours" (5.5).

Understandings of family in the series are thus much in keeping with dominant gay kinship ideologies of the 1980s and 1990s. In "Forever is a Long Time: Romancing the Real in Gay Kinship Ideologies," Kath Weston discusses the impact of both the perceived potential of and active familial rejection on coming out as key influences on gay understandings of family. Such experiences undermine the centrality of dominant ideals of what created "real" kinship by highlighting the fragility of "blood ties"; reformulating notions of kinship along lines of "choice." These "chosen" families, which can include friends, lovers, biogenetic relatives, and ex-lovers in various configurations, are usually framed in terms of tenacity and temporal considerations: authenticity is linked "not only to biology but also to duration" (62), with duration being seen as a combination of chronology and commitment. *Queer as Folk* indeed presents a multiplicity of "families of choice" in different sizes and configurations. We do see families that are nuclear (with a twist), families formed by romantic relationships and sometimes marriage, and both official and unofficial adoptive parenting roles. Perhaps most important for this conception, we see a kin network of friendship in the core friendship group of Brian, Michael, Ted, and Emmett that connects into a larger family that incorporates many of the other main characters in the series, which is indeed seen "as a bond more likely to endure than ties of "biology," marriage, or partnership" (Weston 68).

Yet one can definitely also witness the influence of late twentieth and early twenty-first century discourse surrounding gay family in *Queer as Folk*, for "[a]s gay kinship has become a topic of controversy and debate in society at large, the friends incorporated into chosen families have received little attention, despite the respect historically accorded to friendship by many lesbians and gay men" (Weston 67). The characters in *Queer as Folk* who are most clearly represented as a model gay family of the noughties are Lindsay and Melanie and their children, a family seemingly ready to be strategically deployed in legal and media campaigns for gay and lesbian parenting, marriage, and domestic partnership rights. Their house, their marriage, and (at least for the first two seasons) a stay-at-home mother who asserts that lesbian parents "have to be a little bit better" than heterosexual ones (2.9) presents a certain currently popular ideal of what makes a gay family, perhaps operating as a further textual positioning of "the lesbians" with the assimilationist model.

An emphasis on a more traditional family formation can also be seen in Lindsay's attitudes toward her own family in that she habitually prioritizes and accentuates Brian's primacy as "the father" over Melanie's as coparent, though such attitudes are, at least to some extent, undermined by the series. During the couple's first separation, Brian cruelly taunts Melanie, "I'm his father. [with nasty look] Who are you?" to which Melanie responds, "I may be no one, but at least I love him enough to know that his needs come before mine, which is more than can be said for you" (1.15). Brian to some degree acknowledges Melanie's position by signing over his legal rights to Melanie, stating, "my son deserves two parents who'll be there for him and love him. And who love each other" (1.17). Brian's narrative of growth thus has less to do with taking on the traditional "father" role than acknowledging the lesbian family and recognizing a queer vision of family where love and commitment are of primary importance instead of genetic connection—an idea further seen in his decision to allow Lindsay and Melanie to move to Canada at the end of the series.[24]

Despite these advances and Lindsay's progressive realization that Melanie is her family, it appears that Lindsay's view of parenting is still defined by biogenetic connections. Upon the proposal that Melanie bear their second child, Lindsay insists that Brian once again be the sperm donor, despite the problems they previously encountered and Melanie's great dislike of him. Although Melanie argues that Brian has changed and is "always there to write a cheque," Lindsay's primary argument is that "bottom line—without him, our kids won't be related" (3.3). Instead of attempting to argue that their kids will be related through their upbringing as siblings and with another second-parent adoption, Melanie seems immediately convinced by this latter notion. Brian, having initially agreed, refuses when he learns that Melanie will be the birth mother, and when he changes his mind, Melanie rejects the notion. She argues this to Lindsay on the basis that Brian is "not to be trusted and [is] definitely not the kind of person *I* want as the father of *my* child. . . . You actually expect me to have a baby with someone like that?" Lindsay responds, "I did," and Melanie says, "well, that was your decision. This one's mine, and I say—I want someone else" (3.3). Melanie thereby posits the birth mother as the *real* mother, an assertion quite different from her previous emphasis on parenting as involving love and responsibility but similar to Lindsay's previous insistence on Brian as the sperm donor. The tension between the different and politicized definitions of "family" in these story lines emphasizes the ongoing rhetoric surrounding families and marriage that have dominated the political agenda of the LGBT movement in the noughties.

As we have seen, analyses such as that which Walters undertakes of *Queer as Folk* (and many other texts that present gay families) tend to focus on whether the text truly presents alternative families or simply replicates dominant visions of the heterosexual family with the only change being the substitution of a same-sex partner (which many *do* in fact see as an "alternative" family). The various merits of either side can ultimately be argued with no possible conclusive answer. Weston proposes that "[t]o move beyond oversimplified arguments that 'alternative' families either totally mirror or completely counter 'hegemonic' forms of kinship requires a less dichotomized understanding of the dynamics of social change," as "ostensibly similar formal features of kinship can carry conflicting meanings and embed subtle ideological shifts, allowing "new" family forms to be read simultaneously as radically innovative and thoroughly assimilationist. In the end, they are intrinsically neither." (61, 64)

Rachel Epstein too asserts that "our analysis of lesbian families needs to consider diversity in lesbian experience" (60). Each of these observations is in keeping with arguments made elsewhere in this monograph that declaring a complex representation as either "positive" or "negative" is overly simplistic and that the diversity of lesbian identities, experiences, and desires are more complex than a single representation or politically motivated position will allow.

MAPPING OUT MARRIAGE

Marriage and the "feminine" dramas that go along with it are usually associated with the lesbians of the series, who are referred to by the gay men and Leda alike as "the married munchers." Unlike Michael, who at the end of season four states that marriage "wasn't a story I told myself like straight kids did" (4.13), Lindsay asserts that she has "had the same dream ever since [she was] little—to fall in love, get married, and have a baby" (2.5), locating her as identified with the "straight kids" and idealized notions of what constitutes womanhood. One can witness this identification in both her frequent pretenses of being Brian's wife and her enthusiasm for her imminent marriage to Frenchman Guillaume in the first season. Having broken up with Melanie, she quickly sets up house with Guillaume (who is said to be, but is not entirely depicted as, gay), and the two, under the auspices of "pretending for the immigration department," enact a vigorous reproduction of a heterosexual marriage, replete with a scene in which we see Lindsay trying on a rather ostentatious wedding dress in front of a mirror. Although the threat of this marriage is eliminated by Brian's intervention, this scene acts as a foreshadowing not only of Lindsay's desires for a wedding but also of her yearning for heterosexual attachment.

In the first episode of the following season, we see Lindsay and Melanie both dressed up in frocks and on the arms of "beards" Emmett and Ted at Lindsay's sister's wedding.[25] After many jibes directed at the illegitimacy of Lindsay's partnership with Melanie, Lindsay cracks and proposes to Melanie during the wedding speech. The preparations for the wedding take up a great deal of the couple's story line for the second season, and they finally get married in 2.11. In the lead up to the wedding, in an attempt to gain her parents' acceptance, Lindsay organizes a party (replete with a harp player) to show her parents that she has a "real family with a real life, just like [her] sister, [hoping that then] they might feel differently" (Melanie in 2.7). This attempt at integration is represented as flawed not only because Lindsay's parents refuse to "come to the party" but because it requires closeting, visually represented by Lindsay's breasts: usually braless, her breasts are tightly bound and buttoned for the party, as various characters note throughout the episode. Brian then livens up the party by spiking the punch with ecstasy, and Lindsay eventually takes her bra off and goes downstairs to have fun. Upon prompting from Melanie, Lindsay's parents decide to attend, and by the time they eventually arrive, they witness a roomful of half-naked, dancing queers, as well as their daughter's now unrestrained bosom, foiling Lindsay's attempts at assimilation and prompting her instead to rebut the need for it:

> *Lindsay:* You know the whole point of this party was to prove that we're just like you, so that you'd accept us, the way that I've tried to accept you.
> *Mother:* I didn't realize I needed to be accepted.
> *Lindsay:* You're right mum, you don't, and neither do we. (2.7)

The wedding itself strangely enough has an extremely similar narrative structure to other earlier television gay and lesbian weddings, though with a fascinating inversion. As Walters describes it, "It is interesting to note that in three of the major 'gay weddings' handled on television, it is a heterosexual character who brings the nervous and fighting homosexual couple together when the nuptials are threatened" (*All the Rage* 184). Regarding the 1996 *Friends* wedding of Carol (ex-wife of major character Ross) and her partner Susan in "The One with the Lesbian Wedding" (2.11), Walters comments,

In the *Friends* wedding, Carol rushes to the gang on wedding eve when trouble hits in the form of a fight with fiancée Susan. It seems that Carol's folks are not coming to the wedding and Carol, in her sadness, thinks perhaps they should call it off. But ex Ross comes through and saves the day . . . Both lesbians are thus voiceless, and it takes the wise words of an enlightened heterosexual—comparing it to his own experience—to set the world right. Indeed the last image is of *Ross* and Susan dancing at the wedding together. (184–85)

In *Queer as Folk*, "the lesbian wedding" also appears in episode 2.11 of the season, which is either an astounding coincidence or an indication that the writers of *Queer as Folk* intended the episode as a response to or comment on the earlier rendition of a lesbian wedding on television.

Here, Lindsay calls off the wedding after a comically presented series of disastrous events that lead Lindsay to assert that "our wedding has been methodically and systematically deconstructed. If you ask me, someone up there doesn't think we should get married [Leda sighs]. Maybe there's a reason, maybe that's because people like us aren't supposed to get married" (2.11). After an unsuccessful attempt to convince Lindsay otherwise, Melanie tearfully calls on Brian to fix the situation. Brian, like Ross, has previously had sex with the refusing party, is extremely opposed to the concept of the couple getting married, and is not planning to attend the ceremony, instead choosing to go to a circuit party. Brian draws a comparison between Melanie and "uber-bitch dyke Ripley" from *Alien* before proclaiming, "Christ. Send in a dyke to do a faggot's work!" and assembling his recruits (consisting of gay men and Debbie) to "get the munchers married" because "no one knows wedding shit better than queers" (2.11; again, there is an opposition between dykes and queers).

Consistent misogynist and lesbo-phobic comments are made by all male characters throughout this episode:

Michael: What do you get two dykes who have everything?
Brian: Dykes don't have everything, that's why they're so miserable.
Michael: Well, unfortunately matching penis transplants are a little pricey for a poor shopkeeper like myself.

and

Justin: Lesbians are into endangered species.
Brian: Unfortunately, they aren't one of them.

Melanie, as usual, comes in for particular attack, with Debbie replying to Melanie's insistence that she always does her own hair with "I can tell! [Both Debbie and and Lindsay laugh.] You need more *bella donna*, less Bitchy Butch-y." Walters' critique that in the *Friends* wedding "[t]he straight characters get reformed and redeemed through their expertise in pre-wedding cold feet, thereby avoiding reckoning with their previously impregnable homophobia" (185) seems particularly applicable here, with the "heterosexual" characters simply replaced with gay men and a gay-male-identified straight woman. The episode ends with the bouquet being thrown, and the scene switches to a bouquet (thrown by a drag queen in a wedding dress) being caught by Brian at the white party, clearly announcing the two different types of "white parties" in queer culture: the assimilationist wedding (associated with lesbians, at least until season four) and the queer circuit party (associated with gay men).

In season four, the status of marriage as purely associated with lesbians and straights is somewhat undermined. Ben proposes to his partner (and central character) Michael, and they are legally married during a trip to Toronto (4.13). This spontaneous union (the entire story line from proposal to wedding occurs within a single episode) appears to have little in common with the fuss and drama of Melanie and Lindsay's wedding, which took eleven episodes to develop. However, there are several similarities. Both Michael and Melanie question the notion of marriage before agreeing to their respective proposals. For Melanie, this is due to her view of weddings as "antiquated meaningless rituals for heterosexuals" (2.1), while Michael is hesitant because, as mentioned earlier, he never considered himself able to get married (4.13). When Melanie changes her mind, it is because of "[t]he little girl who asked if [they] were married":

> I realised that some day our kid is going to ask us that very same question, and when he does I have to say no, then he'd want to know why not and then I'd have to tell him because—straight people wouldn't let us. Only that's not the whole truth. . . . We wouldn't give ourselves permission. (2.2)

The doubts these characters experience are further extrapolated by Brian's assertion that "[w]e're queer. We don't need marriage, we don't need the sanction of dickless politicians and pederast priests. We fuck who we want to when we want to—that is our god-given right," to which Michael responds, "But it is also our god-given right to have everything that straight people have. 'Cause we're every bit as much human as they are" (4.13). The gay marriage debate is thus firmly framed in *Queer as Folk* in terms of equal rights, as opposed to the more conservative position espoused by some cultural commentators that it acts as a necessary legitimation or "maturing" of the gay community.[26] Unlike the traditional marital vision of creating a relationship and family, both couples have been together for a significant amount of time when they get married, and both already have a child together. The institution of marriage in *Queer as Folk* is thereby displayed not as an authorizing or legitimizing force but simply as a symbolic affirmation of love. That Michael and Ben's marriage takes place in the very episode in which Lindsay and Melanie break up acts to undermine both the narrative cliché of weddings as a time for reunion and the perception that a marriage is "for life." This is further emphasized by a conversation between Emmett and Ted; the bonds of friendship (sometimes between ex-lovers) are thereby still portrayed as the central organizing unit of family in the GLBT community.

LESBIANS, BISEXUALITY, AND HETEROSEXUAL SEX

The suggestion that Lindsay's lesbianism is somewhat unstable becomes more clearly textual in episode six of the fourth season, when a story line begins to develop that will see Lindsay having sex with a male character by episode ten. This begins via the introduction of the character of artist Sam, who is from the beginning characterized as a difficult and chauvinistic man. Lindsay, attempting to get Sam to show his artwork at her gallery, is initially horrified by both his rudeness and his clear objectification of women. Later Melanie, in a very uncharacteristic move—after listening to Lindsay's ranting that "the man is a pig. . . . If I wasn't already a dyke, he'd have me diving for the nearest muff"[27] and that "he looked through me as if I was invisible, but you should have seen the way he went after his tramp assistant,"—advises her on how to get his attention, and thereby his artwork, following Lindsay's own advice to her in a previous story line: "it isn't right, it isn't fair, it just is. . . . Look, he's a man, you're a woman, gay/straight, it doesn't matter. We all have our powers of persuasion. If you want something out of him, you gotta play his game, 'cause he sure as hell isn't gonna play yours." (4.6).

The next time we see Lindsay, she is clad in over-the-knee black boots and is flirting with and flattering Sam, who proceeds to repeatedly proposition and grope her. After playing along to some extent while attempting to evade his more explicit advances, Lindsay eventually slaps him and storms off. Upon his agreement to show art at the gallery at which Lindsay works, however, she relents, and many scenes of flirtation and jealousy continue as they work on the exhibition together. Sam even inspires Lindsay to once more take up her art. Several conversations occur that use "art" as a barely veiled metaphor for Lindsay's desire for Sam, while her desire for Melanie is expressed in terms of filial connections and responsibilities, acting to rearticulate *Queer as Folk*'s association of creativity and cultural production within the phallic economy, relegating women's relationships and endeavors to the realms of hearth, home, and family.

Lindsay's desires are further indicated by a rare sex scene with Melanie (significantly, their first since episode 2.19) in 4.10. A very pregnant Melanie is performing cunnilingus on Lindsay, who quite extravagantly expresses her boredom with several sighs. Lindsay then requests that Melanie use a vibrator, which, as Sarah Warn remarks, "we're clearly meant to understand is a phallic replacement" ("*Queer as Folk*"). Although the scene soon cuts away, their next conversation informs us that Lindsay was "wild" and "so turned on" (Melanie in 4.10). Later in the same episode, after various displays of guilt-induced domesticity, including not attending the opening of the art show that she has curated, Lindsay has feverish sex with Sam (up against an original painting no less). In a reading of this story line, Bobby Noble argues that *Queer as Folk* herein presents heterosexual sex as more desirable and "as a more creative activity than lesbian sex demonstrating either a backlash logic of reversal (straights as the new marginalized sexuality) or else the queering of queer culture (straight as the new queer)" (149). I believe that there is nothing "new," queer or otherwise, in this story line, as the promotion of heterosexual sex as superior to lesbian sex is endemic to almost all cultural products.

In a review of this series of rather unfortunate events, Warn asserts that the story line "does more damage than good":

> The lesbian-who-sleeps-with-a-man plot device has been used so frequently in entertainment that it dwarfs almost all other representations of lesbians on screen (except perhaps the equally frustrating lesbian-motherhood plot device, which *QAF* also employs regularly), and gives the impression that it occurs more frequently in real life than it actually does—that in fact, *most* lesbians want to sleep with men, rather than just a few. It also makes no distinction between lesbians and bisexual women, and continues to reinforce the notion that bisexual women will inevitably betray lesbians for a man, as if infidelity is the particular vice of one sexual orientation rather than cutting across all. ("*Queer as Folk*")

While Warn's frustrations are completely justified, her later claims that *Queer as Folk*'s Lindsay is among the television characters who "momentarily switch sexual orientations" are somewhat inflated, as Lindsay has been depicted as latently bisexual throughout the series. The surrounding story lines also act to somewhat contextualize, or at least add a degree of complexity to, this plot development. Such narrative contextualization can be seen when Michael and Ben's foster son Hunter is depicted dating a girl, having previously (except in one comment) been presented as gay. In a highly comic series of scenes, Hunter is anxious about coming out to his gay parents as straight, eventually does so, and Michael expresses all the classic parental coming out lines (e.g., questioning if it is "just a phase" and if Hunter just needs to meet the right boy). Ben, in contrast, displays more understanding, although it is an understanding that is underlaid by a polarized view of gay and straight. In answer to Michael's curiosity about how this could be the case considering Hunter's previous sexual experience with men (within the context of sex work) and desire for Brian, Ben reveals that he himself has had sexual experiences with women prior to his coming out. The ongoing insinuations that Brian and Lindsay had sex when they were in college are confirmed in this episode, showing that even Brian, the "ultimate homosexual" of the series, has had a heterosexual encounter.

The clearest narrative correlation is, however, provided by the character of Drew Boyd (Matt Battaglia), who is introduced in the episode immediately preceding the one in which Lindsay sleeps with Sam. Drew repeatedly and firmly asserts his heterosexuality despite seducing and having an ongoing sexual relationship with Emmett. After their second sex scene, they have the following conversation:

> *Emmett:* Does your fiancé know . . . that you're the H word. . . . I meant homosexual, gay, queer.
> *Drew:* I'm not a fag. . . . a fag's sissy, girl, pansy, you think I'm that? . . . A fag can't even throw a ball. You know how far I can throw? . . . I'm a hero to millions. Name one fag that's a hero, that's gotten a call from the president saying "great game." . . . So why would anyone think I was a fag?
> *Emmett:* 'Cause you had your dick up my ass?
> *Drew:* So I like to fuck guys, doesn't mean I love them, wanna kiss them or even know them. It's just for fun, to get off. No one's to ever hear about this, understand? (4.10)

It is noteworthy that later in the season Drew does kiss and indeed pursue Emmett, undermining the veracity of such statements. His speech—reminiscent of the closeted dissociation of homosexual sex from sexual identification in the speech made by Roy Cohn in *Angels in America* (2003)—demonstrates the tenacity and ferocity of internalized homophobia. However, it also acts to raise the issue of whether it is sexual activity or self and/or societal identification that defines the homosexual.

After Drew refuses to come out about their relationship, Emmett pronounces,

> I'm sure most people would agree with you. Why tell anyone? Why lose everything when it can be your little secret. You see, it was different for me. Everyone could tell who I was from the start, which, believe me, didn't make my life any easier. I've been beaten up, cursed at, spit on from day one. Which, in a way, was worth it. Because I've never had to live a lie. And I'm not about to start now.

This monologue emphasizes the importance of coming out and engaging with the experiences of visible homosexuals. During the fifth season, however, Drew *does* come out as gay, and his narrative potential to be a bisexual character is thereby removed. Both the content and ferocity of Emmett's sentiment is echoed in Melanie's outburst:

> *Melanie:* I know which team I play on. It's not a choice or a preference. It's who I am. It's who I've always been. A rug muncher, a muff diver, a cunt lapper, a bull, a lezzie, a dyke.
> *Lindsay:* What do you think I am?
> *Melanie:* Don't ask me to make up your mind for you. You have to do that all by yourself.
> *Lindsay:* I'm a lesbian.
> *Melanie:* Not if you're having sex with a man, honey. (4.12)

The language used here is significant. Whereas the terms Melanie uses to describe her sexuality are either signifiers of cunnilingus or reclaimed pejoratives (at least one of which refers to the ultimate visible, stigmatized lesbian), Lindsay responds by identifying with the somewhat more polite term "lesbian."

Even this self-naming Lindsay undertakes is actually somewhat unusual for her character. While the male characters frequently refer to themselves as fags and Melanie refers to herself as a dyke, Lindsay usually elides naming herself, instead using the term "you know" or in other ways skirting around linguistic definition. This can be seen either as part of Lindsay's closetedness or in terms of a certain discomfort with her seemingly homosexual self-definition. Melanie's assertion that "[i]t's not a choice or a preference" further distances her sexuality from Lindsay's, as Lindsay has repeatedly framed her lesbianism after her encounter with Sam in terms of choice and preference (e.g., "I still choose [Melanie]" [4.11]). Barbara Ponse's distinctions between "*elective* lesbians who experience a great deal of flexibility and fluidity in terms of sexual object choice, and *primary* lesbians, who see their sexual possibilities as much more restricted and exclusively directed toward the same sex" (paraphrased in Stein, *Sex and Sensibility* 17) appear rather pertinent to the distinctions between Melanie's and Lindsay's sexual identity.

So it seems *Queer as Folk* is here attempting to undertake an examination of what constitutes homosexuality, where the boundaries are, and what it takes to be queer. Such an emphasis is heightened by the text's refusal to name bisexuality. Sexual experiences with the opposite sex for "homosexuals" or with the same sex for "heterosexuals" are seen as aberrations to whom one "really is" (Ben in 4.11), and the *verbal* suggestion of bisexuality is never made. Even Lindsay, who has had an ongoing struggle between her desires for Melanie and for Brian and in season four clearly is conflicted between her desires for Sam and for Melanie, does not question whether or not she may be bisexual. This reluctance to name is telling because it speaks to the definitional reformulation of the term "queer" by *Queer as Folk*. It is not clear whether these story lines are intended to open up *Queer as Folk*'s formulations of sexual identity or return them to a fixed state. During Lindsay's confession to Brian about her affair, he tells her "it's okay to like cock, and it's okay to like pussy, just not at the same time" (4.11). Lindsay herself also consistently expresses her sexuality in terms of an either/or choice. When Sam asks her, "what about the part of you that secretly yearns for something else? The part of you that both of us know is there?" Lindsay responds, "my house has many rooms, I occupy but a few, the rest go unvisited" (4.11).

Even at this juncture, however, *Queer as Folk* passes on having Lindsay come out as bisexual or at least narratively questioning her sexual orientation. Her character's attitude for the last few episodes of the season was unlikely to garner much audience sympathy: she is not terribly understanding of Melanie's anger (although she could perhaps relate it to her own earlier experience as the "betrayed" party), presumes that Melanie has forgiven her after a couple of days, and then fluctuates between being a doormat laced with self-pity and having flares of self-righteous anger at Melanie. Perhaps if she felt confused or generally repentant or even broke up with Melanie, she would have appeared as a more sympathetic character in these episodes. And I suspect that was the point, for in *Queer as Folk* the Kinsey scale is generally closer to a traffic light without the orange, despite the many heterosexual-identified men who sleep with men. The emphasis placed upon "choosing a team" by both Melanie and Brian appears to be one that the series as a whole espouses.

BUTCH-FEMME CODINGS IN *QUEER AS FOLK*

It is of definite significance that it is Melanie who is depicted as clearly lesbian while Lindsay displays attractions to men. From sexological discourse to pop-cultural representations and even within queer culture itself, there are often distinctions drawn between the "real" or "authentic" lesbians, who are almost always portrayed as masculine or at least more masculine than their consorts, and those whose sexual preference is depicted as more fluid, who are generally marked by their femininity. *Queer as Folk*'s Lindsay is portrayed as "less lesbian" and more identified with men than her partner Melanie precisely because she is coded as the femme. In this way, the series activates a very traditional view of feminine lesbians as latent heterosexuals. While "lesbian chic" may have already seen two blooms, little it seems has changed in the last two decades in regards to how feminine lesbians are represented in popular culture. This observation must be tempered by the fact that the vast majority of lesbians who appear in popular discourse at all *are* feminine in both appearance and narrative identification. The point remains, however, that the sexualities of those characters *identified by the narrative with masculinity* (whether or not they actually visually express female masculinity) are not similarly positioned as associated with fluidity of sexual object choice. I have discussed various factors that call into question the stability of Lindsay's self-identification as a lesbian: her attitude to and relationship with Brian, her frequent lack of apparent lesbian sexual desires, her sporadic closetedness, and most obviously her sexual encounter with Sam. I believe that the text presents these factors as inextricably concatenated with her femininity.

Writing about the early lesbian romance film *Desert Hearts*, Mandy Merck comments that "[t]he brunette/blond: active/passive dichotomy is now an established convention of the lesbian romance" (379). This dichotomy can also be observed in *Queer as Folk* and in *The L Word* as discussed in Chapter 4. Merck also observes that *Desert Hearts* doesn't invoke the literal, commercially off-putting codes of butch and femme, but it does employ traditionally related dichotomies of class (cheeky casino girl pursues shy professor), geography (candid Westerner courts aloof Easterner), sexual history (experienced lesbian brings out previously faithful wife), and appearance (passionate brunette warms up cool blond, who honors a long cinematic tradition by eventually letting her hair down). (379)

Queer as Folk in contrast does invoke codes of butch and femme, albeit in a subtextual manner (presumably also to avoid being commercially off-putting), and these codes are deeply entwined in the narrative with the other "traditionally related dichotomies" that Merck discusses: class (while Melanie is an attorney and Lindsay a teacher and later a gallery curator, Melanie's mostly nonprofit work and Lindsay's conspicuously monied background render them within traditional codes), geography (Melanie is from New York, while Lindsay is from the more conservative Pennsylvania), sexual history (although Lindsay makes one verbal allusion to a past girlfriend, her primary previous sexual attachment is figured as Brian throughout the series, while Melanie's previous romantic attachment is Leda and Melanie frequently makes reference to having "always been a dyke"), and appearance (passionate and feisty kind-of-butch brunette warms up prudish and reserved blond, who only really lets her hair down for a disheveled and chauvinist male artist). These dichotomies are deeply entangled with understandings of "active" masculinity and "passive" femininity invoked by traditional models of heterosexuality and furthered by intellectual traditions such as psychoanalysis.

In *Queer as Folk*, one cannot help but notice the curious practice of *encoding* Melanie as butch while not *presenting* her as such, much as characters were once coded as queer while not giving any definitive visible indication of their queerness (most famously discussed by Vito Russo in *The Celluloid Closet*). Her appearance is less feminine than Lindsay's, with her fairly short, dark hair, less voluptuous physique, and usual garb of pants and shirts. But she is certainly still visibly within acceptable codes of femininity: her hair is not *too* short, she is conventionally beautiful, and she wears makeup and even occasionally wears dresses for work or social occasions. Her encoding as butch instead operates through contrast and on a verbal level. Such characterizations usually take place negatively, via jokes or insults at Melanie's expense on the part of the gay male characters (and at times Debbie), though they are sometimes more positively framed by Melanie herself or, on a couple of occasions, other characters. This need to encode rather than show is particularly peculiar in light of the portrayal of Emmett as a flamboyantly effeminate gay man, pointing toward the manner in which male femininity has been commodified (and thus rendered acceptable) in a way that female masculinity has not.

Despite this lack of visual female masculinity, in many of the comments made by characters in *Queer as Folk* there is a definite conflation between lesbianism and masculinity (e.g., Michael's assertion that Gus is "being raised by two lesbians, [so] he's going to need a feminine influence" 1.6) and butchness/lesbianism and lack of style (e.g., "less bitchy butch-y" in 2.11 or Ted's comment that flannel is "lesbian lingerie" in 1.21). The presence of Melanie allows *Queer as Folk* to project its cacophony of misogyny at a particular figure without potentially offending their targeted audience of straight women and without having to actually represent any of the subjects of their phobias, thereby keeping them in sublimated form: female masculinity, feminism, or inattention to the whims of fashion are only present in the subtext of the series. One of the few times that a woman who is clearly physically expressive of female masculinity is given a foregrounded role (still of course a one-liner), she is characterized by the narrative as grotesque and fear-inspiring. Attending Gus' bris (1.3), Michael is greeted by a larger butch woman with a "hi Michael" and a hug. While this hug lasts for the briefest of moments, Michael displays an expression of extreme distaste and acts as if he is being squashed.

Whether or not Melanie actually visually performs "butch," the traditional associations directed at butch-femme relationships are present throughout the series, although these associations have been gradually undermined within lesbian cultures over the last two decades. In *Queer as Folk*, the butch is seen as the authentic lesbian, who is figured as the desiring party with a higher sex drive than the femme, and the sexual top. Melanie also prioritizes her bond and attachments to women and is at times portrayed as anti-men. While *Queer as Folk* is willing to present a small spectrum of differently gendered presentations with its gay male characters (contrast, for instance, the campy Emmett and the hypermasculine Brian) without defining one as inherently "more gay" than the other, "true" lesbianism appears to only be accessible within the series to those in possession of (butch) lesbian signifiers.

What does this portrayal say about the depiction of lesbianism in contemporary culture? *Queer as Folk* has certainly been groundbreaking for its medium in its presentation of gay and lesbian characters in a queer world, in its candid sexuality, and in its willingness to engage with contemporary political issues that directly pertain to homosexuals (e.g., same-sex marriage, gay parenting/adoption/custody, assimilationism, gaybashing, state clampdowns on sexuality, and so on). Yet the manner in which its lesbian characters are represented has seen little change from earlier depictions of lesbianism. Lindsay, when now given the choice, may choose Melanie instead of Sam, but the instability of her identity formation—seen not only through this encounter but also in her relationship with Brian—has much in common with the ongoing tendency to categorize feminine "inverts" as wayward heterosexuals without true sexual desire toward other women. Melanie is still subject to verbal abuse on the basis of her unfeminine qualities and is frequently betrayed by her partner's prioritization of male characters. Even Leda (codified as promiscuous) becomes irreversibly clingy once she has slept with Melanie and Lindsay. And while they may be more sexual than the lesbians previously seen on television, they are significantly less sexual than other characters in their textual world. *Queer as Folk* reprojects the image of the loathsome homosexual—whether in the form of unfeminine lesbians or gay, conservative rhetoricians—rather than attempting to remove the image altogether. Perhaps the clearest lesson coming out of *Queer as Folk* is that even when operating within a queer framework, attention still needs to be paid to the specificities and privileges of gender. Those who would seek to explain some of these failings through the largely gay male production team and audience may be interested to know that the more clearly lesbian text *The L Word* shares several of these aspects with *Queer as Folk*, particularly the representation of femininity, as will be elucidated in the following chapter.

CHAPTER 4

RECYCLING *THE L WORD*

MUCH HAS BEEN MADE OF HOW "DIFFERENTLY" THE TELEVISION SERIES *The L Word* (2004–) represents lesbianism. The hype surrounding *The L Word* purports that the series ushers in a new era of lesbian visibility and representation to the mainstream, which presents a fashionable and glamorous image of lesbianism to counter "the stereotype" of lesbianism in a curious repetition of the popularised notion of late 1980s and early 1990s "lesbian chic." During this period, the visible lesbian subject is claimed to have shifted from the "mannish lesbian" of modernity to the decidedly more marketable "lipstick lesbian." As Martha Gever remarks in her monograph on lesbian celebrity: "If understated mannish garments and bearing could be said to constitute lesbian visibility in the past, the 1990s witnessed the arrival of a lesbian style that is decidedly more spectacular and, as a result, feminized if not always conventionally feminine—flashy but not necessarily frilly." (39)

This vision of lesbian style is displayed in *The L Word*, and the marked lesbian body is given significantly less representational prominence. It is even at times explicitly disavowed, most obviously in Alice's disparaging reference to what she terms a "hundred footer"—"Is it her hair? Is it her jog bra? Is it her mandles?"—remarking that she can "tell she's a lesbo from across a football field" (1.11). The primary focus of debate over *The L Word* has been whether these "femininized if not always conventionally feminine" (Gever 39) images can be subversive or not.[1] In this chapter, I would like to move beyond that somewhat polarized debate and instead perform a through exegesis of how *The L Word* represents and constructs the lesbian subject. Unlike most examples of lesbian representation on television, the ensemble nature of *The L Word* allows for multiple and differential constructions of lesbians, even if this multiplicity is revealed primarily through contrast and within a fairly glamorous, standardized spectrum. As Eve Kosofsky Sedgwick put it, in *The L Word*, "[t]he sense of the lesbian individual, isolated or coupled, scandalous, scrutinized, staggering under her representational burden, gives way to the vaster livelier potential of a lesbian ecology" (B10). This lesbian ecology does not spring full-grown from the heads of creators Ilene Chaiken et al., despite what celebratory media reports might insinuate, but rather recirculates a multiplicity of understandings (and clichés) of the lesbian subject and lesbian life.

Investigating Femininity: Lipstick and Lesbians

The manner in which the femme is theoretically visualized has seen much change in the last twenty years. This owes much to the reclamation of butch-femme cultures engendered by the sex wars, of course, but also, as Danae Clark has asserted, "some credit for the changing perspectives on fashion might also be given to the recent emphasis on masquerade and fabrication in feminist criticism and to the more prominent role of camp in lesbian criticism" (190). These and other influences resulted in an intensified critical engagement with the figure of the femme, enabling her to be seen as something greater than a capitulation to and replication of heterosexual and misogynist norms. Such influential essays as Sue-Ellen Case's "Towards a Butch-Femme Aesthetic," for example, focus on the camp masquerade of the femme, whose femininity is performed to the butch, which Case suggests points toward a new possibility for a feminist subject position. However, even in this account, which seeks to resignify the femme as a positive and authentically lesbian subject, she is only visible, and seemingly only *possible*, in the presence of the desiring, or in some views the desired, butch.[2] Other writers, notably writing from a more personal and historical perspective, such as Amber Hollibaugh, Cherríe Moraga, and Joan Nestle, have articulated both lesbian authenticity and agency in the femme, rearticulating as active and subjective modes of desire previously understood as passive and reclaiming objectification and receptivity as potentially powerful.

Clark's "Commodity Lesbianism" and Arlene Stein's "All Dressed Up, but No Place to Go? Style Wars and the New Lesbianism" discuss the then-emerging phenomenon of lesbian chic, attempting to ascertain the affirmative and injurious possibilities of this newfound (hyper)visibility. Later accounts, such as those of Sherrie Inness and Ann Ciasullo, have further read the significance of 1990s images of lesbians in mainstream discourse. Others, notably Martin and Walker, have focused on the manner in which the feminine lesbian is presented in contemporary theory and literary practice. It is into the context of these obviously intersecting yet rarely integrated traditions of reclaiming and theorizing femme identities and cultures (within specifically lesbian contexts), as well as the context of gay and lesbian studies' readings of feminine lesbians in mainstream culture that I am attempting to locate my reading of *The L Word*.

The growth of images of "lipstick lesbianism" can be seen as both a by-product of mainstream media attention, which inevitably favors a "consumable lesbian" (Ciasullo 584), and the explosion of femme theory and writing in the 1990s, which could be characterized as the decade in which "the femmes . . . [were] finally asserting themselves" (Clark 199). This complicated confluence of factors, which could be endlessly reconstituted in terms of cause and effect depending on one's perspective and desired outcome, is usually rather simplistically characterized as either the "discovery" of lesbianism by the mainstream or the "rediscovery" of our "real," playful selves emerging from under the haunting specter of lesbian-feminism. Gever has noted that lesbian celebrities must posit "a measure of the acquiescence to gender and sexual norms required for recognition and inclusion to occur peaceably" (43). Likewise, in *The L Word* we have images that have been constructed, at least in part, for a heterosexual media and populace, and these images embody such measures of acquiescence for economic and political purposes of peaceable inclusion and integration—an embodiment that is considered to be particularly necessary for the televisual medium that displays them.

This chapter takes up the central issue of visibility and the relationship between the visible and the real. *The L Word* has an odd approach to and relationship with visibility, simultaneously desiring and disparaging it, announcing and obscuring it. It inserts the lesbian into a frame in which she is to this day fairly invisible, notably utilizing a lesbian image that is historically invisible from the cultural imagination of what constitutes a lesbian via her frequent indistinguishability from heterosexuality. In doing so, the visible lesbian is rendered invisible through closeting, the invisible (though not "real") lesbian is rendered hypervisible, and lesbianism itself is rendered simultaneously more and less visible.[3] The "visible difference" situated in the butch lesbian (Martin 105) has enabled her positioning as the "magical sign" of lesbianism (Walker, *Looking Like What You Are* 201) expressed in multiple texts, from *The Well of Loneliness* to lesbian theory to *Queer as Folk*, as the "real" or authentic lesbian. It is perceived that because the marked (butch) body is "already and always marked as lesbian, she is *more* visible than the femme—and thus, if represented, more 'lesbian' than the femme" (Ciasullo 601–2). The formulation that the femme is popular culture's visible lesbian and yet by virtue of her "sameness" simultaneously invisible—hence unable to adequately represent lesbians—is a fairly common one. The converse of such arguments views the femme's "likeness to" (sometimes problematically expressed as proximity to) heterosexuality as rendering her uniquely able to denaturalize heterosexual perceptions of lesbianism and, by extension, heterosexuality itself. The publicity materials surrounding the series further add to this perception, as the theme of such materials seems to be to distance the series from "old-style" lesbians—characterized as those who are a little too feminist or a little too butch.

Considering the outward "reclamation" of femininity undertaken by *The L Word*, it is curious that in the first and rare appearance of the actual *term* "femme" in the series, it is used as a synonym for heterosexual. The second episode of the series, "Let's Do It," features a lengthy sequence in which the protagonists engage in a mission to find out whether or not the woman (Lara) that Dana (Erin Daniels) is interested in is a lesbian. They observe Lara's (Lauren Lee Smith) garb, movements, and reaction to Bette and Tina kissing, utilizing the pseudoscientific methods of placing her various attributes in "lez" and "straight" columns and assigning them points:

Bette: Well, she's got some good lezzie points for her walk, and the way she moves that chopping knife.
Shane: Yeah, but she's way femme-y on the coiffure tip.

It seems significant that here it is not "girly" or "feminine" that is used as a term that is interchangeable with heterosexual, or at least a marker thereof, but "femmey"—a term seldom used outside Western lesbian subcultures, which denotes a particular style of lesbian. It is quite mystifying that in a series that centers on a group of predominantly "lipstick lesbians," "lezzie" and "femmey" can be placed in such marked opposition to one another. While visual and attitudinal identification of lesbians as lesbians often, if not usually, relies on recourse to signifiers of the marked butch lesbian, one would think that this group of predominately "femmey" dykes would be aware of other codes or would at least problematize this statement, pointing out that feminine presentations do not necessarily denote unequivocal straightness.

In the final scene of the "Let's Do It" story line, Lara passionately kisses Dana in the locker room. Her consequent statement, "just in case you were still wondering," undermines the analysis of Lara as straight based on visible signifiers portrayed as definitive by all the characters, reaffirming that a lesbian does not have to be visible as such to be one. This functions as almost a cautionary tale about using butch signifiers as the magical signs of lesbian gaydar, allowing for a demonstration of the elasticity of lesbian signifiers and an affirmation that sexual identity cannot always be read from the body or its ornaments. Despite this message, however, there are further implications to the episode "mission: gaydar" than simply undermining the preconceptions of the characters and audience as to "what a lesbian looks like." These scenes highlight a key facet of *The L Word* that is primarily found in the deeper recesses of the text—that "real" lesbianism is to be found in (implied) female masculinity—and rupture this message through to the surface in the form of dialogue.

Walker's assertion that "the feminine lesbian cannot be studied in isolation from the idiom of race passing" (*Looking Like What You Are* 11) is of particular resonance to an examination of *The L Word* due to its presentation of and engagement with both the feminine "passing" lesbian and the light-skinned "passing" African American. The issue of passing is further explicitly raised in the show via an altercation between Bette and Yolanda (Kim Hawthorne), an African American, single mother-to-be in Bette and Tina's therapy group for prospective parents. Whereas in the previous example, Lara's passing is not an attempt or even a desire to pass but rather a figural tendency to do so, here the question of intentional passing, or what is perceived to be intentional passing, and the ethics thereof is raised. Yolanda, who has already been characterized as "confrontational" in the episode, criticizes Bette for emphasizing her lesbianism while never referring to herself as an African American woman, asserting that "you need to reflect on what it is you're saying to the world while hiding so behind the lightness of your skin" (1.8). This discussion is ended later in the episode when Bette outs Yolanda as a lesbian. Bette accuses Yolanda of practicing a double standard by allowing herself to be mistaken for a straight woman, saying that "you're not exactly readable as a lesbian, and you didn't come out and declare yourself." The writers' aim in constructing this improbable and quite melodramatic scenario is perhaps to emphasize that one should not judge those who appear to pass, as it is not necessarily possible to announce all identities through bodily markers or overt declaration.[4] When Bette asks, "why is it so wrong for me to move more freely in the world just because my appearance doesn't automatically announce who I am?" (1.8) it further questions the centrality of visibility to truth, perhaps attempting to counter the emphasis placed on coming or being out by gay and lesbian political groups and individuals. Bette's statement, however, also problematically insinuates that light-skinned biracial persons, or feminine lesbians (as the two are somewhat conflated in this story), "pass" because they wish to "move more freely in the world"—that the unmarked body connotes a simple relationship of conforming or desiring to conform. The appearance of the unmarked body is therefore figured as a performance rather than a complex expression of identities.

The L Word, while ostensibly advancing the cause of feminine lesbians, appears to reinforce perceptions of the feminine lesbian as inauthentic or, at the very least, to express a deeply ambivalent attitude toward lesbian femininity. Even as it overtly disparages non-feminine women as unfashionable, it is when Dana attempts to dress in a *more feminine* manner that she is disparaged as unfashionable and "geeky" (1.3). When (hyper)femininity is associated with lesbian tennis player Dana, it is seen as a restrictive force, symbolic of her closetedness—a mask used to demonstrate heterosexuality. The two most significant instances thereof are the dresses she wears to Bette and Tina's sperm-hunting party in the pilot and when attending her first Subaru party in "Lawfully" (1.5), which are associated with, and emblematic of, her attempts to pass as heterosexual and in marked opposition to both her usual casual sporty clothes and the tailored pantsuits she tends to wear after coming out. In the first episode, her floral patterned, empire-line dress is seen as part of her heterosexual drag, available for activation (along with her doubles partner and beard, Harrison [Landy Cannon]) should any heterosexual people show up at the party. In "Lawfully," her pale pink, extremely tight dress is held together by ribbon lacing all the way up the back. This is of particular interest because in this scene she is taking Harrison out as her official "date" instead of her girlfriend Lara, who comes over to Dana's apartment, assuming that she had been invited. Having an openly lesbian and very sexual partner makes her attempts to pass more difficult on a literal and emotional level and requires more restriction—here presented as more femininity: she is quite literally restrained by pale pink ribbon and, unused and unsuited to such constriction, she can move only her arms and take very small jerky steps with her corseted chest heaving like a lady from a Regency romance.

There are moments in *The L Word*, however, that do function to undermine this proposition. In answer to the question of what she will wear on her first date with Lara, Dana proposes a "blue sundress" to howls of disapproval from her friends. Her answer to this chorus of negativity is "but I'm going to a nice place, y'know, somebody might see me" (1.3), stressing both her closetedness and the ability of femininity to disguise obvious lesbianism. This association is almost critiqued once again by Lara, who is wearing a dress and initially looks at Dana with horror, feeling overdressed in comparison to Dana's (supposedly casual) outfit of a white, fitted button-up shirt and pants. Lara asks if she is thereby "a geek" for having worn a dress, and Dana responds, "no, I'm a geek. For letting my friends tell me what to wear" (1.3). This could be read as an indication that her friends' constant attempts to make her dress more androgynously are a source of pressure on her and that she really *does* feel comfortable in her dresses. The fact that (at least for the rest of the first season) her dresses seem to disappear immediately after she does come out indicates that this is not the case; when she is now required to "dress up" for formal occasions or simply going out, she eschews dresses in favor of pantsuits or jeans with a simple shirt. During season two, when further emphasis was placed on displaying designer dresses (even struggling-to-make-ends-meet waitress Jenny appears in new designer coats and dresses every week) and under the influence of fiancée Tonya, Dana once again begins to wear dresses, but that does not negate the very clear narrative message of the first season.[5]

In season two, Jenny (Mia Kirshner), as part of her continued coming out and claiming of her lesbian identity, engages in the classic trope of such claiming: cutting off her previously long hair ("Lynch Pin" 2.4). This move is precipitated by a discussion she has with new (straight) housemate Mark (Eric Lively) about whether or not she visibly appears to be a lesbian. When questioned as to why he can identify the other on-screen lesbian characters as such but not Jenny, Mark tries to explain: "It's not that they're masculine, or anything, 'cause actually some of them are pretty feminine. You know? It's . . . they have these . . . haircuts. These very cool haircuts—don't get me wrong—it's not—more—it's obviously more than a haircut. But it's—no, it's true. It's this . . . something that they exude that's . . . (*thinks*) I'm gonna try and put my finger on it." (2.4)

Although initially expressing disbelief at the limitations of this assessment, Jenny appears to take to heart this appraisal of visible lesbianism as being the possession of a cool haircut—or at least seems to desire to render her sexuality more visible—as later that night she asks Shane to give her a haircut, saying that "I just feel like I . . . need to change" (2.4). The single tear rolling down Jenny's face as Shane cuts her hair off is reminiscent of her mourning for heterosexual identity in the death of her alter ego Sarah Schuster. The next time we see Jenny, she is strutting down the street alongside Shane with her new short haircut, looking far more confident than she has before and even being cruised by a female passerby as a song plays, whose lyrics repeat "butch in the streets, femme in the sheets" ("Labyrinth" 2.5). This moment is perhaps a reversal of the Dana story line, as here Jenny must take on a (slightly) more butch persona, at least "in the streets," to both assert her lesbianism and become empowered while perhaps her femme side "in the sheets" is portrayed as the deeper, more authentic level of her personality.

The L Word seeks to distance the lesbians of the series from "old style" lesbians—no matter how feminine or palatable to a mainstream audience they may be. The series predominantly, but not solely, associates its feminine lesbians with the "normal" lesbian, the "lesbians' blank page," and places this in opposition to the term "femme," particularly in their manifestations as "scarlet starlets" (Blackman and Perry 74, 68) with their celebration and camping of femininity. As Lesléa Newman notes, "Even in the gay nineties, with lipstick lesbians reigning supreme, some women [found] it an insult to be called a femme" (11). And, for many femmes, including those who enjoy their lipstick, "lipstick lesbian" is "a derogatory term that conjures up an apolitical creature . . . a lesbian who doesn't want to be a dyke and doesn't want to be associated with dykes" (Walker, *Looking Like What You Are* 211). Despite such antagonisms between the two, they are often conflated, particularly within mainstream discourse. Alexis Jetter claimed in a 1993 *Vogue* article that "[p]utting [k.d.] lang aside," lesbians in the mainstream "have a few things in common: They're white. They're middle class. And they seem more interested in makeup and clothes than in feminism. In short, they're femmes, or what the straight world prefers to call lipstick lesbians" (88). Academic discourse is also not unwilling to assert the same point; Ciasullo, for example, cites Jetter's conflation unremarked and repeats this fusion between femme and lipstick lesbian throughout her argument. The characterization of lesbians in late twentieth-century mainstream culture that Jetter describes is (unfortunately) completely true. The subsequent conflation of the rich cultural history of queer femme identities with the culturally and temporally specific, consumerist, and invented-by/for-the-mainstream image of "lipstick lesbianism" is a vast oversimplification, although distinguishing them from one another on the basis of *image* alone is difficult if not impossible. Jetter's formulation, repeated in the discourse pervasively, completely annexes the central role in femme history of femmes of color, working class femmes, and feminist femmes.

Such characterization of feminine lesbians also cycles back to previous modalities of lesbian representation and noticeably the first major appearance of lesbians in U.S. popular cultural discourse: lesbian pulp novels of the 1950s and 1960s. In these texts, in light of prohibitions that insisted that the characters could not end up happily together, it was more often the more feminine lesbian who would end the story in a heterosexual relationship, abandoning and at times vigorously rejecting her lesbian desires. *The L Word*'s nod to 1950s butch-femme culture takes place in "Liberally" (1.10), which sees Dana and Jenny meeting by chance in a bar, which the newly bisexual Jenny in wonderment observes is "like something out of the 1950s. It's so butch and femme." The newly out Dana replies despondently that they indeed are in "the oldest lesbian bar in L.A. Actually, it probably hasn't changed *since* the 1950s. But really, it's no different than any other club, you know, I mean, you have a few drinks, and you talk to a few people you have nothing in common with, and realize how unlikely it is you'll ever meet anyone who's right for you again."

Later in the episode they are suddenly and somewhat inexplicably shown at Jenny's abode, and what then proceeds to occur is surely one of the most awkward sexual encounters in the history of television. Although Dana is, as Walker would say, "your jock type, not your butch type" (*Looking Like What You Are* xiii), in these scenes the discussed context of the bar, together with the contrast between Dana's outfit (she is clad in a white singlet and jeans), mannerisms, and attitude and those of Jenny (dressed in a black dress, pink sash, and silver heels), indicate that Dana is intended to be read as "the butch" in this scene.

The awkwardness of the ensuing attempted sex act is engendered by the fact that neither of them appears to know what to do to the other.

This is not simply a matter of the sexual inexperience of both parties but is because they are both sexual "bottoms" (or at least have been portrayed as such at this point of the narrative—Dana later becomes a "top" in her second-season relationship with Alice). The scene acts to challenge perceptions of visible difference as sexual difference and to undermine conceptions of the butch as necessarily a top, a theme gradually being promulgated in lesbian culture but entirely apparent neither in *The L Word* nor in other mainstream depictions of lesbianism. There are many ways this scene can be read. However, a defining sensation thereof is that both parties, and particularly Jenny, are *play-acting* desire. This is expressed by such factors as Jenny's big, pseudo-interested eyes and comments like "wow! That's really interesting," when "that" clearly is not interesting and her words contrast with the monotone in which they are uttered. This is problematic and troubling, particularly considering that in one of the few other times butch-femme relationships are explicitly discussed (although butch-femme codes are in almost constant implicit use in the series) it is referred to as "role play" (1.13 and 3.3). One cannot help but wonder if *The L Word* is attempting to suggest that such identities are simply imitative and constricting "roles" that do not work. This is further explored in season four via Shane's relationship with single mother Paige Sobel (Kristanna Loken). Toward the end of the season, Shane imagines herself and Paige within a perfect vision of 1950s-style nuclear family domesticity (replete with appropriate costuming and sets) in a series of fantasy sequences. The parodic nature of these sequences and their representations of butch-femme "roles" acts to situate butch-femme as a throwback to an earlier era and to critique these identities as old-fashioned and conservative.

It appears that many cultural theorists who seek to view the femme as positive do so in light of the figure's supposed proximity to heterosexuality. Inness, for example, in the first chapter of her book *The Lesbian Menace: Ideology, Identity and the Representation of Lesbian Life* reads two 1920s texts, one that presents a "mythic mannish lesbian" and one that depicts a feminine lesbian. Inness reads the representation of the feminine lesbian as revealing "a greater threat to heterosexual order than does the mannish lesbian" by virtue of her inability to be visually distinguished from heterosexual women, as the femme destabilizes "an image of the lesbian as an easily excluded outsider" (31, 32). When it comes to lesbians in later mainstream discourse, however, even Inness appears to shift somewhat on the point of how radical the feminine lesbian can be, noting that "[a]lthough I agree with Clark and Groocock that lipstick lesbians are too complicated to be viewed as merely a sign of lesbian cooption, I am far more ambivalent about how representations of lipstick lesbians or 'designer dykes' are manipulated in the mainstream press" (74).[6] Throughout Ciasullo's survey of various magazine and filmic depictions of lesbianism in "Making Her (In)visible: Cultural Representations of Lesbianism and the Lesbian Body in the 1990s," she finds the central trope to be that of the "consumable" lesbian: "Mainstream culture is thus giving with one hand and taking back with the other: it makes room for positive representations of lesbianism, but the lesbian it chooses as 'representative,' decoupled from the butch that would more clearly signify lesbianism for mainstream audiences, in effect becomes a nonlesbian" (599–600). Ciasullo later expounds this perception in relation to *The L Word* in an interview conducted by Booth Moore for a newspaper article on the series:

> Most people envision lesbians as butch dykes in sleeveless flannel shirts and jeans—so how to represent lesbians on TV in a politically correct way becomes a quandary. "Think about images of African Americans, and someone like Sidney Poitier, who was seen as changing the image of black men in film but by some critics was seen as an Uncle Tom figure," Ciasullo says. "The same thing goes for lesbians. The stereotype is the butch lesbian, and to get away from that, you have the feminine lesbian. But as images get feminized, lesbianism gets subsumed." (Moore and Ciasullo qtd. in Moore)

I am rather discomfited by the analysis that "as images get feminized, lesbianism gets subsumed." I am concerned here that "[i]n our efforts to challenge forms of gender policing, we run the risk of replicating a kind of gender totalitarianism, even in the form of its deconstruction" (Martin 119). Theorists should be wary of replicating the too easy association of femininity with heterosexuality, even within the loaded discourse of popular culture.

Yet there is much truth in the analysis of the lesbian representations seen in popular culture as attempts to airbrush the specter of homosexuality through the disavowal of those that visibly or sexually disrupt normative cultural precepts. In her analysis of 1990s women's magazines during the first blush of lesbian chic, Inness observes that "[r]eaders are encouraged to look at the stylish, surface appearance of lesbianism, not to seek beneath the surface for a deeper understanding" (63). Similarly, *The L Word*—or if not the text itself, then at least the discourse surrounding it—focuses on style and stylishness: surfaces. The deeper layers of signification are not only consequently overlooked but also frequently, in fact, activated to counter the meanings produced by the surfaces they have a hand in representing. This is particularly seen in the casting of at least outwardly heterosexual actresses in the central roles, with the exception of one out lesbian actress, who is notably cast as bisexual. Once again, this resonates with the earlier depictions, which used "models who look[ed] stereotypically heterosexual pretending to be lesbians" to provide "titillation without threat as there is an implicit understanding that these are not 'real' lesbians" (Inness 65–66).

While such writers as Erin Douglas feel that the sexuality of the actresses in *The L Word* is not relevant to their performance as lesbians ("Femme Fem(me)ininities"), the seemingly deliberate casting of mostly heterosexual actresses has various repercussions on the signification of lesbianism that the series produces—perhaps most significantly in reinforcing hegemonic perceptions of lesbian sexuality as a liminal, mobile state easily returned to heterosexuality via the lesbian performances of straight actresses who then, in interviews, are at pains to focus discussion as much as possible on their husbands and boyfriends. As much as one might insist on the text itself as the only producer of meaning, it does not exist in a vacuum, and multiple factors, including previous knowledge of the actresses and perhaps most particularly the media flurry accompanying the program, can come into play in a viewer's reading of the series. This is to some degree countered by various insinuations and, at times, statements that the cast may not be as straight as they seem.[7] Whether or not this is true—which it very well may be, as evidenced by Jennifer Beals's (Bette) seemingly accidental outing of Katherine Moenning (Shane) in early 2006 (see Stockwell)—one cannot help but feel that this practice replicates the simultaneous giving and withholding of what Clark refers to as "gay window advertising," wherein "lesbians can read into an ad certain subtextual elements" (183). In *The L Word*, the audience can read "real" closeted lesbianism into the actresses, thus reading themselves into the frame and rendering *The L Word* "a model of the new gay marketing strategy" in that they are "believable as lesbians to lesbians—but just barely" (Phelan 197). As much as scorn has been heaped on the inauthenticity of the feminine cast, *The L Word* has often also been praised for that very feature, as it is seen to resignify the sign "lesbian" from its popular associations with female masculinity and thereby demonstrate that there are feminine lesbians (or, as some have asserted, that lesbians are "just like" heterosexual women). If *The L Word* is indeed to show its audience that there *are* feminine lesbians, however, is it not then a trifle curious to do so by using mainly heterosexual women to *perform* lesbian femininity?

Walker, in her historical study of the feminine lesbian in literature and theory, found that

> [i]n general, these women are coded traditionally as either the "wifely partner," with a woman's financial and emotional dependence and a burning desire to darn socks, and/or as the wayward heterosexual who returns to men when "the life" becomes too difficult (when the going gets tough, the femmes go straight). (*Looking Like What You Are* 43)

Feminine-coded characters are consistently associated with vanity and narcissism in the text, as well as (often sexual) duplicity. The depiction of feminine lesbian desire as less authentic or somehow less dedicated to women (seemingly by virtue of their femininity) can also be seen in *The L Word*. Several examples of just such a "return to men" occur in the series—the most significant of which is Tina's desire for men and subsequent relationship with Henry (Steven Eckholdt) in seasons three and four.

During the third season, we see Tina experiencing desire for men while back in a (sexless) relationship with Bette and subsequently acting on those desires. From the outset of the series, Tina's maternal qualities, together with her long-term monogamy, low sex drive, and dedication to Bette's career, act to situate her as the "wifely partner" whom Walker describes: a perfect model of a 1950s housewife, a woman who "not only mimes but embodies a version of traditional femininity" (Walker, *Looking Like What You Are* 53). It is significant, then, that Tina Kennard goes from a long-term relationship with Bette Porter (Jennifer Beals) in season one to a relationship with Helena Peabody (Rachel Shelley) in season two before developing a seemingly overpowering desire for men. Whole scenes are devoted to her obsessing over men and making out with a male employee before getting involved in a relationship with another man and even placing a cybersex advertisement that reads "dyke w/baby seeks real man for good fuck . . . slide your big cock into my blonde pussy" (3.5)—a statement completely in keeping with heterosexual male pornography's fantastical depictions of lesbians. Although Tina's "return to men" is explained away by the character in the fourth season as being accounted for by her "humiliation at finding out that her lover was cheating on her" (4.9), this does not narratively make sense in light of both her relationship with Helena that directly followed Bette's affair or the intensity of her desires for men as they were portrayed in the third season, which occur on a purely physical rather than psychical plane. While Tina's desire for men wanes in the fourth season, with Henry being characterized as "boring" (Tina in 4.11), the fact remains that this story line is in keeping with the narrative traditions of many pulp novels wherein male characters "always [got] the girl—at least, the most feminine girl—in the end" (Keller 3). Perhaps this is because mainstream society finds the idea of the feminine lesbian—who appears to be visually indistinguishable from heterosexual women—threatening, and portraying these lesbians as always potentially bisexual undermines the challenge of depicting such characters, assuages the discomfort of potentially homophobic viewers, and allows male viewers to feel that they "have a chance" with such women. As Samuel A. Chambers observes, "The pilot has plenty of sex, just not very much between two women. . . . Eventually, of course, *The L Word* offers numerous and repeated portrayals of lesbian sex. . . . Yet reminders of the implicit message that lesbians are sexy, attractive objects of desire, even for straight men, crop up repeatedly" (90–91).

It should be noted here that the model of femininity embodied by Tina is accessible to her via her very specific class positioning. Her financially privileged position renders her the mistress of a large house with a referenced but invisible housekeeper, and she is only in need of a second income after the breakup of her marriage. This is in keeping with a more general trend in depictions of lesbians on television in recent times that creates images that hearken back to earlier, "purer" times. Television acts to disassociate these characters from connections between lesbianism and masculinity or "sexual perversion," enabling a proposition of lesbians as "just like" heterosexual women or perhaps as even more perfect examples of traditional "womanliness" than contemporary heterosexual female television characters. As Dana Heller argues, the lesbian couple on *E.R.* "[is] portrayed as more family oriented than any of the heterosexual characters on the show" ("States of Emergency" 29), and this also seems to be the case with Tina in *The L Word*, as her total absorption in prospective motherhood (her entire season one story line revolves around it) renders her more clearly within the parameters of a conservative feminine ideal than most contemporary examples of heterosexual female characters on television. In *The L Word*, however, due to the multiple lesbian and bisexual characters, the pressure on a single lesbian or lesbian couple to signify many aspects of lesbian life is not as great, so the baby story line does not have the same desexualizing effect as it does on, say, *Queer as Folk*.

Interestingly, unlike triumphant narratives involving lesbian conception that act as a positive affirmation of "alternative families," such as *If These Walls Could Talk 2*, *The L Word*'s lesbian childbearing ("gayby") story line is quite differently coded, for while the series opens with the joyful "let's make a baby," the insemination and pregnancy are fraught with difficulty and heartache. This could be seen as suggestive that Tina and Bette's aspirations to this model of traditionalized heteronormative life with its firmly ascribed roles (one earns the money, the other picks up the dry cleaning) are potentially and ultimately destructive to the selfhood of individuals and thereby relationships.[8] Or perhaps it is simply an element of the narrative drive toward tragedy in *The L Word*. Douglas notes that "[a]s the show progresses, Tina's character seems to get stronger as she steps farther away from her 'authentic' feminine performance [as housewife]" ("Femme Fem(me)ininities" 55). This is then built on during the second season, wherein Tina is once again pregnant and, despite certain attempts at manipulation by equally dominant women, is suddenly and rather inexplicably as a result of her separation with Bette in possession of emotional, sexual, and even physical power that she had heretofore never appeared to possess.[9] After having the baby, the couple's relationship resumes, but they seem unable to consummate it, and Tina uses the child to "pick up" divorced father Henry—motherhood in *The L Word*, it seems, is not conducive to lesbian relationships, much as it was not in *Queer as Folk*.[10]

DESIRE AND RELATIONSHIPS IN THE L WORLD

Lesbian desire is frequently figured as structured alternately by similarity or by complementary difference. In *The L Word*, both of these tropes are utilized in presenting the couplings of the series: those relationships that display codes of "difference" are characterized as more effective or possible (though ultimately these too are ineffective and come undone), while the relationships characterized by similarities—visual, racial, or interests-based—are exemplified by unrestrainable desire and beguilement. The primary lesbian relationship in season one of *The L Word* is portrayed as being structured by visual and narrative gendered and racial difference, and it is (arguably) these differences that lead to the disintegration of Tina and Bette's seven-year relationship. Bette is the wage earner, more commanding and involved in a very demanding career, while Tina is depicted as the "wifely" partner, staying at home to "prepare [her] body for pregnancy" (1.1). The visual association of femmes in general—and filmic/televisual lesbians who wish to bear children in particular—with blondeness is reinforced by this representation, which mobilizes cultural associations of "natural" femininity with whiteness. Their difference is prominently visually coded through their clothing styles; Bette generally wears designer suits with tailored men's shirts—these being marks of a necessitated corporate femininity—while Tina predominantly wears casual "peasant-style" clothes in keeping with her confinement to the sphere of the home. Even when both the characters are attired in suits, they are generally, and quite amusingly, put in one white suit with a black shirt and one black suit with a white shirt (almost as if one is a photographic negative of the other), so although Bette and Tina are dressed almost identically, they are still coded as opposite and complementary to one another. Their emotional and communicative styles are also very different: while Bette does not enjoy verbalizing her feelings, Tina is very enthusiastic about (and loquacious at) therapy, which seems to render them both completely unable to communicate with one another.

This lack of communication extends to the couple's sex life, which has been less than perfect for three years out of their seven-year relationship. During the first season, the first two of the three Bette/Tina sex scenes depicted are either for the purposes of, or inspired by an attempt at, procreation. The third occurs in "Limb from Limb" after Tina has discovered Bette's affair. This scene—one of the few incidences of physical violence in the series—is instigated through the "out of control" anger of the usually fairly passive Tina, who slaps Bette, and Bette's response of restraining her and attempting to have nonconsensual sex with her. Tina regains control of the situation, seizes Bette's hand, puts it inside herself, proceeds to use Bette's hand to satisfy herself, then collapses on top of her and the scene ends. This scene is not only very confronting but also rather perplexing in terms of its signification. Perhaps it is the culmination of Tina's recurrent casting as the victimized wife, and her victory in this sexually violent power struggle is symbolic of Tina's subsequent move away from Bette (the final scene of the season sees Tina at Alice's, distraught and minus her wedding band). The violence of this scene—or perhaps, rather, the violence of domestication—is suggestively prefigured by the presence of Catherine Opie's photographic *Self-Portrait* (1993) in both the (flashback) scene in which Bette and Tina meet (1.11) and, more lingeringly, at the gallery opening at which Tina discovers Bette's infidelity (1.13). The portrait depicts a naked back on which a picture is engraved. The razor engraving on the skin is a childlike drawing of a house, a cloud, and two girls (delineated as such via their triangular skirts) holding hands. The violence of this image's inscription at the beginning and (for the time being) the end of their relationship suggests that the childlike dream of perfect lesbian domestic felicity in a traditionalized mode depicted by this image and aspired to in Bette and Tina's relationship requires a certain degree of violence to the self, and Bette and Tina's violent battle for (sexual) dominance in this scene acts to bring the submerged power relations of such a heteronormative relationship to the surface.

Candace's (Ion Overman) behavior, race, and gender presentation are visually presented as entirely different from Tina. Her most marked characteristics are toughness, assertiveness, and desire, together with an impermeability and impenetrability directly contrasted to Tina's softness and fluid boundaries between self and other. These features are reflected in her profession (carpenter), which requires both physical strength and manual skill. Candace is often seen in overalls or singlets and pants: practical, butchy clothes that leave her upper arms bare to highlight her strength. Although she has long hair, she keeps it effortlessly slicked back into a ponytail. In the two scenes in which she is not dressed for work, she displays a somewhat more femme veneer, with more makeup and somewhat more feminine tops; however, the fact that her outfits are in the same style as her usual wear (sleeveless tops with pants), combined with her various mannerisms and the contrast between Candace and the other lesbians seen in proximity to her in these scenes, still render a certain butch persona. Candace not only aggressively pursues the object of her desires but also takes control from Bette during sex, who has been hitherto seen as both a top and a control freak. Candace is also figured in relation to Bette's African American heritage;[11] Bette meets Candace through Yolanda, and they first meet at one of Kit's performances, which is the first scene in which we see a large group of people composed primarily of people of color. Interestingly, Bette first looks at Candace with explicit desire after Slim Daddy (Snoop Dogg) expresses his captivation with Candace and with the idea of her and Bette being together, which perhaps functions to render this narratively illicit desire as being accessed through the male gaze. Douglas argues that "Candace offers Bette some type of authentic racial performance. Their attraction is depicted as almost carnal. Bette's desire is a need to not only consume Candace but also her race as well, which is very problematic" ("Femme Fem(me)ininities" 58). Whether or not this is the case, it is certainly clear that Candace is more like Bette than Tina is, and it is these similarities that appear to draw Bette to her, seemingly almost against Bette's own will.

The other key example of unrestrainable and illicit desire is also structured by similarity as opposed to difference. From the first, Marina and Jenny's similarities are foregrounded—from their interests and tastes, to their hair color and complexion. That the zenith of their figural correspondence is achieved at the peak of their relationship indicates the significance of notions of similarity to the representation of their relationship. This scene takes place in the short time period between Jenny's separation from Tim and knowledge of Marina's relationship with Francesca and immediately after a scene in which we see a previously distraught and extremely dirty Jenny curled fetus-style in Marina's bath. The scene displays the only time they openly go out together as a couple, and they are dressed in identically fawn-colored shirts, with Marina in black pants and Jenny in a black skirt. This lesbian party signals Jenny's "birth" into a new, lesbian world, into which Marina will shortly abandon Jenny, having "opened up [her] world" (Marina in 1.7). Such an awakening is, however, depicted as figuratively impossible without a death. This allegorical death takes place in the form of Jenny's (heterosexual) "fictional" alter ego Sarah Schuster who drowns at *sea* (the symbolic and maternally coded *Marina*) shortly after she consummates her relationship with Marina (1.2). Together, these metaphoric associations form an expression of Lacanian discourse that "theorizes homosexuality as a desire to return to the moment of primary identification, and lesbianism in particular [as evocative of] that primordial signifier of mirroring, the mother-child dyad" (Walker, *Looking Like What You Are* 142).[12]

However, not only psychoanalytic but sexological discourse is recalled by the recurrent allusions to lesbianism as vampirism, a "fantastic . . . demon possession sort of thing" (Jenny in 1.4). Metaphors of vampirism and contagion hold sway in Jenny's seduction by Marina: from their first encounter—in which the camera pans from Marina's hypnotic stare, to Jenny's transfixed one, to extreme close-up shots of their mouths—to the first time they have sex, which is presented as both exquisite pleasure and extreme loss and pain.[13] Here, the falsely human visage of the vampire that allows it entry into human environments is replaced by that of femininity, which allows the "falsely heterosexual" visage of the feminine lesbian unsuspicious entry into the heterosexual world. With her lesbian identity "obscured by the 'mask' or 'cover' of friendship," the feminine woman "can invade even that site of heterosexual sanctity, the home" (Inness 30), thereby providing her with further opportunities for seduction of "heterosexual" women. This is emphasized through Tim's jealous conviction that Jenny is sleeping with a male friend, which ironically almost leads him to discover her and Marina in *flagrante delicto*. The multiplicity/unplaceability of Marina's aristocratic cultural background (she speaks multiple "foreign tongues" in her language of seduction) further contributes a vampiric lineage to her depiction. In *Reading the Vampire*, Ken Gelder discusses national identity in relation to *Dracula* and the discourse surrounding it, locating polyphonic abilities and mixed lineages within the characters of Vambery and Dracula. This is then read as an anxiety about reverse colonization, loss of national identity, and the ability to traverse national boundaries (11–13), which can be extrapolated in the case of *The L Word* as an anxiety about the bisexual/homosexual woman colonizing the heterosexual woman into a state wherein the stability of her sexual boundaries become fluid.[14]

FEMALE MASCULINITY?

Shane is the most cogent example of *The L Word*'s rendition of a butch figure in the group. Despite her much-discussed lack of short hair in season one, Shane's clothes, walk, posture, and mannerisms, together with the contrast provided by the other characters, all embody butchness, rendering her the most visible lesbian of the protagonists—as Dana says to her in the pilot episode, "every single thing about the way you're dressed, like, screams dyke" (1.1). This is seen as integral to her character, as she does not, like Candace, "scrub up femme" for big events. This is despite her job as a hairdresser, an industry not generally associated with butch women and one in which her clients are depicted as primarily (at least outwardly) straight women. Shane is also extraordinarily stylish, and her very androgyny is portrayed as chic. Despite the aspersions cast by Alice in "Looking Back" on "the hundred footer" (a lesbian easily identifiable as such from a hundred feet away) in the narrative, here is a hundred (or at least a fifty) footer who is not marked by the series as undesirable; indeed, she is the much mooted "lothario" of the narrative. Ciasullo, in her analysis of 1990s mainstream images, asserts that mainstream discourse has two quite contrasting ways of representing butch women. She can either be depicted as "masculine and *undesirable*" (601) or, like Gina Gershon in *Bound*, for example, be depicted as both butch and "simultaneously marked as feminine with her pouty, Julia Roberts lips, wispy hair hanging in her eyes, and her reputation as an actress" (589). Shane appears to have much in common with Ciasullo's latter characterization—she quite literally has the "wispy hair hanging in her eyes." As far as her reputation as an actress is concerned, Moenning is not, unlike Gershon, usually dressed as a feminine character, previously having played both a transgendered character and a cross-dressing teenager. Unlike Hillary Swank, for whom "[m]edia coverage of [her] nomination and subsequent selection by the Academy [for playing transgendered Brandon Teena in *Boys Don't Cry*] emphasized her 'real-life' femininity in contrast to the boyish Brandon she played on screen" (Surkan 167), Moenning is promoted via her *actual* likeness to Shane's androgynous or tomboyish characteristics, thereby presenting a less clear-cut vision of "femininity restored." [15]

The emphasis placed on the very desirability of Shane, both within *The L Word* and in interviews and reviews, is a significant one. It appears that co-writer and producer Chaiken, in particular, makes a conscious effort to uncouple the signifiers of "butch" and "undesirable" in mainstream culture. Chaiken declares that Shane/Moenning: "brings that revolutionary androgyny that confounds. She can pass for a boy, *yet* she's totally sexy. I think *men respond to her* as much as women do" (in Shister; my emphasis). The repeated use of the word "yet" in articles regarding Moenning and the character she plays is of particular interest because it acts as both an apology and a celebration. The most apparent insinuations here, of course, are that women who can pass for men are not usually "totally sexy" and that it is necessary for *men* to be attracted to her in order for her to be attractive. Despite the fact that within the narrative Shane is primarily attractive *to* women and that these women are gay and "straight" alike (as Tina says in "Let's Do It," "the Shane test pretty much works on every woman"), the emphasis placed on her attractiveness to gay men (in the series) and straight male viewers is quite disconcerting in its need to "validate" the masculine woman through the male gaze.

Here we return to Inness and her assertion regarding the enforcement of "the idea of the 'correct' lesbian being a consumer and a style maven" (Inness 75) in 1980s and 1990s magazines. *The L Word* formulates a vision of the butch as a "correct lesbian" in these terms—both consummate consumer and style maven. Shane's stylishness despite, or rather because of, her look that "screams dyke" (Dana 1.1) is consistently emphasized. She is also emblematic of the idea of the "lesbian as style maven" (a concept quite rabidly promoted in 2004 in tandem with the release of *The L Word*)[16] through her abilities as a hairstylist. Indeed, her lover Cherie's (Rosanna Arquette) husband Steve (James Purcell) appears to envision Shane as a kind of reverse *Queer Eye for the Straight Guy*—a lesbian who can make straight women look hot. Her ability to do so is predicated on her difference, much as the presenters of *Queer Eye* trade on their campy femininity to lend an edge to straight (and it is thereby presumed masculine) men; that is, they teach/groom them how to perform for their partners. Here the (supposedly) straight woman can be taught/groomed to look effortless, nonperformative, far from the primping and, it is insinuated, the "artifice" of femininity. The fact that Cherie is given a tousled "just fucked" hairstyle (i.e., a really messy one) is what makes her look hot. Steve thinks Shane "could be a gold mine" (1.9) and hence invests in her salon because he believes this is a style or a mode that can be traded on, giving women "looks" that show that they appear not to care, even though these "looks" are in fact highly stylized. The joke is, of course, that Cherie's hair has not in fact been styled, nor is it the style itself that imparts sexiness upon Cherie. She sports a real "just [almost] fucked" look as opposed to a synthetic performance of such. That Shane loses her salon as quickly as she acquires it can be seen as recognition of the vagaries of the acceptance of homosexuality as style, fad, novelty, and service provider.

Shane and Cherie's relationship is perhaps the most obvious expression of butch-femme styles in the series and the mobilization of figuring lesbian desire as structured by difference. Cherie is a seducing, active bottom, displaying herself to be taken, and hence she plays with and confuses notions of active and passive sexuality in the manner suggestive of the "actively orchestrating" femme described by such theorists as Hollibaugh (Hollibaugh and Moraga 408). The power of seduction and the power of withholding are firmly placed in the hands of Cherie, and in these hands the previously uninterested-in-relationships Shane is transformed into the well-known figure of the wounded and manipulated butch. Starting out in the series as an almost compulsive bedhopper, it soon becomes clear that such activities are for Shane not merely a love of sex and women nor an indication of sex-positive queer culture but related to her depiction as emotionally stone and deeply wounded. Martin has suggested that "[l]esbian butchness always seems to emerge in the form of a wound or woundedness" (40), and Shane's depiction is in keeping with the "melancholic loner image, which resonates with a whole history of butch representation" (Halberstam, *Female Masculinity* 227). The culturally inscribed association between masculinity in women and melancholy is not a new phenomenon, and its presence in *The L Word* evokes the specters of psychoanalytic and sexological discourses of inversion and the representation of melancholy butches in such classic lesbian narratives as Radclyffe Hall's *The Well of Loneliness* or Leslie Feinberg's *Stone Butch Blues*.

Shane's womanizing is depicted as lack, her desires fulfilled yet unfulfillable. The articulation of desire as lack is a common theme in *The L*

Word and, I conjecture, even a guiding force to the narrative drives of the text.[17] Marina explicitly outlines this notion in the final episode of the first season as part of her attempted seduction of Jenny's lover Robin: "The Greek word, *eros*, denotes want, lack. The desire for that which is missing. The lover wants what it does not have. It is by definition impossible for him to have what he wants, if, as soon as it is had, it is no longer wanted" (1.12), quoting from a text by Anne Carson that she describes as "very romantic." Although this is the case with many of the series' relationships, this pattern appears to find its ultimate embodiment in the character of Shane, for whom "the psychoanalytic notion that all desire is founded in lack seems to solidify in relation to the stone butch as true lack, as real castration, and as the exact place where, to paraphrase Marilyn Hacker, lust tumbles into grief." (Halberstam, *Female Masculinity* 112)[18] This is marked via the repetitious nature of her one-night stands, which are figurally associated with the repetitious nature of her drug habit. A further example of links drawn between butchness and melancholy in the series can be found in the character Lacey (Tammy Lynn Michaels), who appears for three episodes as Shane's stalker. Lacey is the only foregrounded female character with short hair during the first season, and she wears other accoutrements of butchness. Lacey's obsession with Shane's abandonment of her turns out to be transference/displacement/projection of her feelings of familial abandonment, abandonment that Shane too has experienced. Her fixation on an inappropriate (unresponsive) object of desire and her consequent grief is also remarkably similar to what Shane experiences later in the season with Cherie.

By the end of the first season, Shane has been fragmented by love, fragmented by a femme who will ultimately always choose the societal/economic comforts of heterosexuality over her lesbian desires. This fragmentation is most markedly performed in the *mise-en-scène* of their final breakup, which takes place at an art gallery. The scene opens with Shane and an artwork of impressions of body parts imprinted in what appears to be blood, framed by a mirror that is in itself surrounded by (or framed within) pink and blue neon tube lights. She then enters the hall of mirrors proper, which contains Cherie. Throughout their interaction, their images are linked, separated, multiplied, distorted, confused, and fragmented via the presence of the multiple mirrors. At the beginning of this scene, the camera journeys through what appears to be a mirror into the "real" world. Mirrors thenceforth operate as a symbolic representation of the realms of both possibility and delusion. The scene uses a "deep surface camera technique"—a technique that, according to photographer Del LaGrace Volcano, uses mirrors to "reveal both the back and front of the body," which has the effect of making "the viewer feel that they are seeing beyond the surface when, in fact, we are just seeing more surface" (Volcano and Halberstam 39). Due to the multiple mirrors in this scene, there are even further surfaces shown; however, it is difficult to see the surfaces themselves. The last image we see of Shane in this first season is of her sitting in her truck outside of what was to become her salon, on the front window of which appears the half-scratched out sign that once bore the store name "Shane"—a metaphoric obliteration of her selfhood.

The second season sees Shane butching up somewhat. She finally gets a short(er) haircut and frequently wears slightly less androgynous outfits of ties and jackets. Shane's personal evolution in this season sees her fall in love and reject it before slowly opening herself up to it, and she ends the second season on a happy note. Her professional advance is once again offered then removed due to a powerfully controlling woman's interest in her—the vital shift in this season being that Shane actively rejects being controlled. In "Loyal," Shane visits a confessional to expurgate that "[e]veryone . . . wants something from me, and . . . I don't feel like I have anything left to give," specifying that sex is "mainly what people want" from her (2.8). This confession can be seen as indicative of *The L Word*'s mirroring of Butler's perception of stone butches "whereby that 'providingness' turns to a self-sacrifice, which implicates her in the most ancient trap of feminine self-abnegation" ("Imitation and Gender Insubordination" 25), an argument that Judith Halberstam critiques (*Female Masculinity* 127–28).

The character of Ivan (Kelly Lynch) in the first season appears to further the representation of female masculinity in *The L Word* and could be said to be the first serious attempt to queer binary notions of gender in the series.[19] Ivan is first introduced in "Locked Up" (1.12) through a drag act to the song "Savoir Faire," in which he wears a velvet suit and an elaborate pompadour in a parody of and homage to 1970s masculinity (specifically the new-wave writer of the song, Willy/Mink DeVille). The increasing popularity of drag kinging, particularly in lesbian communities, and post-Butlerian notions of gender performativity inform this double presentation, which functions both as an embodied sense of butch/genderqueer/transgender masculinity and a campy deconstruction/celebration of it via drag. Reading Ivan in terms of the drag king categories proposed by Halberstam, Ivan is situated most clearly in relation to the categories of butch realness and denaturalized masculinity in that "the category of butch realness is situated on the sometimes vague boundary between transgender and butch definition" within which "masculinity is neither assimilated into maleness nor opposed to it; rather it involves an active disidentification with dominant forms of masculinity, which are subsequently recycled into alternative masculinities" (Halberstam, *Female Masculinity* 248).

During and after his first drag act, Ivan and Kit flirt and talk and consequently arrange to meet for coffee the following day. Upon entering the café, Kit does not recognize Ivan, who is out of drag and now wearing a leather jacket, jeans, a white button-down shirt, and a large belt buckle and sporting "a long mane of hair—styled somewhere between a quiff and a mullet" (Farrar). Although he is initially unrecognizable to Kit, there are strong visual and behavioral linkages between the two gender presentations that are suggestive of his drag act being explorative rather than transformative (Halberstam and Volcano 131). There is much debate both on the show and among viewers as to Ivan's "actual" gender identification. The character only gives us one indication thereof: toward the end of "Locked Up," Bette corrects Kit on her reference to Ivan as "he," and when Kit apologizes for her "mistake," Ivan states that he is "happy either way." An exposition of Ivan's gender identity is most immediately narratively pertinent in terms of Kit's sexual identity. Bette's warnings to Kit that Ivan is "courting you old school" and "*she* wants to be your husband" lead Kit to question her (previously seemingly unquestioned, or at least not greatly so) perceptions of Ivan as male. This questioning then leads to Kit clarifying to Ivan that she is a "two-months-from-50-year lifetime heterosexual woman" and telling him, "If you were a man—you would be the perfect man" but that, under the circumstances, a relationship between the two of them would not work out. The stability of the identity categories of gay/straight and man/woman is questioned by Ivan's reply:

Ivan: Do you know what you're looking for, Kit?
Kit No. No, not in the big picture sense that you mean.
Ivan: Then how do you know I can't give it to you? (1.13)

While there is a tendency on the part of critics to wish to claim Ivan as either butch or transgender, undoubtedly to broaden the representational diversity of the show, the character resists definition.

Ivan appears to be an elaborate and somewhat convoluted pastiche of a variety of people and personas—storyteller Ivan E. Coyote, Heather Spear's drag persona The Gentleman King, and Willy "Mink" DeVille. Ivan's hair, like Shane's, has been the subject of much discussion, both due to its proximity to the dreaded mullet and its length, which acts to feminize Ivan's appearance. This can be read as a manifestation of Gever's "measures of acquiescence" (43) discussed at the outset of this chapter, with the creators of *The L Word* perhaps perceiving a butch or transgender drag king who could genuinely pass as male to be too threatening for their desired audience. Interestingly, however, both Ivan's wig for drag performances and his everyday hair appear to be inspired by two of DeVille's hairstyles—perhaps dearticulating the connection between long hair and femininity and demonstrating Ivan's portrayal of alternative rather than normative masculinity. The main inspiration for Ivan, suggested by Chaiken herself,[20] is Coyote, a Canadian "writer, storyteller, tin whistler, lighting technician and performer" (Taste This 221). According to Coyote,

> [I have been] trying to define my gender my whole life, and I'm beyond it now.... I don't care about labels or pronouns, I don't identify with "he" or "she." But I don't really like the term "trans-gendered" either; it sounds like you're moving from one state to another. I just am who I am. (Coyote interviewed in Parks)

Considering the character Ivan in light of Coyote can conceivably here lighten the pressure on Ivan to signify a particular gendered identity, instead allowing him or her to become a character who refuses gendered signification, who is not "reducible to transsexual man, transgendered man, or stone butch" (Halberstam and Hale 285).

Spear, in her incarnation as The Gentleman King, performed Leonard Cohen's "I'm Your Man" during the debut of the midwestern king troupe Dykes Do Drag in 1999 (Surkan 173–76), a drag act that is repeated in *The L Word* and acts as the climax of both Ivan's performance of male gender and his attempted seduction of Kit. The usage of this song creates "an interesting juxtaposition between Spear's androgynous look and Cohen's ultra-masculine baritone voice" (Surkan 174). In an interview, Spear remarks that she selected Cohen's song to "position the singer as a versatile subject . . . as a woman I can be that man" (Spear qtd. in Surkan 174). Cohen's song is used by Ivan as a further response to Kit's assertion that "if you were a man, you would be the perfect man" (1.12), declaring that not only *is* he a man but that given the chance, he will be the perfect man for her. This scene complicates notions of gender and what it takes to "be a man," while the lyrics simultaneously perform a deconstruction of classic models of manhood by articulating masculinity as capable of versatility and passivity. Although the act is ostensibly *about* the interplay between performance and "realness" in constructions of maleness, reactions to Ivan's performance generally hinge upon notions of authenticity. The lyric "if you want another kind of love, I'll wear a mask for you" comes under particular scrutiny (see, for example, Lo "It's All About the Hair"), as it is seen to imply that Ivan is "really" a lesbian forced to wear a "mask" of maleness to achieve her desires. In contrast, it can be viewed as a concurrent assertion and subversion of "true" maleness if gender is seen as always a mask or if, as Douglas sees it, the act introduces at least the "concept of gender performance to mainstream audiences/dominant culture that still might see gender as essential and tied to biological sex" ("Femme Fem(me)ininities" 62).

The second season of *The L Word*, despite the producers' promises to be more daring in the series' images of nonnormative gender presentations, did not appear to seriously attempt to play out the character of Ivan or his story line with Kit. In the first episode of the second season, Ivan continues to romance Kit, including giving her the keys to his apartment, and she appears to be falling for his charms and/or attentions. When Kit uses these keys and accidentally witnesses Ivan half-dressed, Ivan becomes extremely distressed and angry, shoving her out of the room and later being unwilling to see her. In "Lap Dance" (2.2), Kit manages, with some difficulty, to locate Ivan, during which time we see her refer to Ivan as "she" for the first time. Kit makes little attempt to reconcile with him or discuss the incident; instead, the major reason for her effort to locate him is her desire to ask him for investment money rather than an endeavor to resolve the situation between them, a request for money to which he eventually acquiesces. Ivan does not reappear in the series, nor is he discussed, with the exception of a brief encounter in episode nine, when Kit goes to him after being stood up by her current (married) beau, and it is revealed that Ivan has been seeing Iris, a burlesque dancer (who interestingly is depicted as having a very low opinion of lesbians, suggesting that her relationship with Ivan is indeed a heterosexual one), for five years and that monogamy "just doesn't work" for Ivan. That he had not previously explained this to Kit, nor disclosed his relationship with Iris while attempting to romance Kit, implicates Ivan as, if not dishonest, at least not straightforward, and therefore he is easily narratively dismissed as a potential suitor for Kit. Ivan's presence in the series, like that of Lisa the lesbian-identified man, appears to have been included more as a sort of gender-bending sideshow than a serious attempt to use the character to engage in discussion of diverse or multiple gender (and sexual) identities.

Before the beginning of the third season of *The L Word*, the inclusion of a new character played by Daniela Sea was vigorously promoted in the lesbian media. The series, criticized for focusing "on the experience of gender conforming lesbians" (Moore and Schilt 160), positioned this character as a response to such criticisms, implying that Sea's character would "bring a butch sensibility to the show that has been criticized for playing it too safe on the genderqueer spectrum" (Lo, "Interview").

However, it was not long before it became apparent that *The L Word* intended to use Sea's character as a counter to more than one critique of their representational strategies, selling the character as both a butch woman and a transgender man, which resulted in disastrous portrayals of both, reinforcing old stereotypes that butch lesbians are or want to be men, as well as presenting female-to-male transsexuals as some sort of gender traitors—assimilationist, misogynist, and eager to embrace "male privilege." This is perhaps because, as becomes clear through the conversation had by many of the established characters toward the end of the episode "Lobster" (3.3), the creators of *The L Word* offer little complexity in their depiction of gender and desire and view butch and femme identities as some sort of heterosexual "role-play" (1.13 and 3.3). While butch characters did appear in lesbian pulp fiction, the most famous of whom is undoubtedly Beebo Brinker from Ann Bannon's novels, in the *visual* popular cultural medium of television, the butch lesbian poses an even greater threat as her visible dissimilarity with normative conventions of femininity calls into question the naturalized relationship between women and "feminine beauty" that advertisers are so keen to promote. As Malinda Lo adroitly observes in an article about the character's narrative arc,

> It's too bad that as soon as that "real butch" sauntered onto the scene, she transitioned from female to male in a clumsy storyline that reduced the complexity of transgender issues to a stereotypical war between the sexes. To make matters worse, Moira's transition into Max was written in a way that not only dismissed the possibility of butch identity, it ridiculed it. . . . What is disappointing about this engagement with masculinity is that *The L Word* is a *lesbian* show. By only allowing men—or women who are in the process of becoming men—to display or engage in masculine behaviors or attitudes, *The L Word* continues to deny a major part of what has

made lesbian cultures so fascinating and so queer for hundreds of years. ("Gender Trouble")

That is, *The L Word* once again hesitates at presenting something that may challenge conventional notions of gender; presenting images that are mediated for, and directed toward, a presumed heterosexual audience.

Season four sees *The L Word* attempting to make up for some of their more irresponsible portrayals of transgendered men in season three, most clearly through the discussion of Max's taking of illegally obtained hormones in a medically unsupervised manner at a transgender support group meeting that plays something akin to a poorly produced and acted public service announcement. However, Max's representation is still clearly a mediated one, whether this is framed through questioning the "reason" behind his transsexuality (4.8) or through the visual construction of this character. As Candace Moore and Kristen Schilt have noted of Shane's androgyny, Max's masculinity too

> lacks full believability on-screen. . . . This allows the show's producers to have their cake and eat it too: they are able to successfully introduce the notion of a woman being read or intentionally passing as male without visually alienating squeamish viewers by rendering one of *The L Word*'s permanent characters male in appearance, or worse, gender ambiguous. (161)

Max's girlish voice, floppy hair, and tiny soul patch do little to convince the viewer that his conservative boss and coworkers would truly accept him as male. No mention is made of the difficulties transitioning people experience in terms of name and gender changes on official documents and bank accounts (that one would assume would have had to be presented for his employment) or of the fact that the "male privilege" that Max is depicted as willingly entering into is indeed not accessible for many female-to-male transsexuals.

As if in deference to the "squeamish" viewers Moore and Schilt identify who may find transgenderism difficult to take, the episode "Lez Girls" (4.5) opens with a scene of Max undressing in front of a mirror. This scene could have been used to great effect to undermine conventional and ideologically ascribed notions of sex as "naturally" dimorphic by demonstrating the real-life masculinizing effects that hormone therapy can have on an ostensibly "female" body. Indeed, if *The L Word* had wished to make an attempt at realism in its depiction of a transgendered man who has undergone a breast reduction and hormone treatment, they could have used digital technology to superimpose Sea's head onto the body of a real-life transman. Instead, we see Sea's clearly female body, and the camera focuses on a hairless chest (which is unlikely given the amount of testosterone Max is reputed to take), breasts, and even freshly waxed legs. This narratively unnecessary scene seems to deliberately reaffirm for viewers who are skeptical of his transsexuality that Max "really" is a woman," and this acknowledgement of essentialised "womaness" is implied as necessary for him to defend a female colleague as he does later in the episode. Once again, this can be seen as a mediated representation, one in which the depiction of a transgendered man can only take place in this popular cultural format if it is seen to offer a voyeuristic "insight" and play into dominant conventions of gendered representation.

Art, Pornography, and Representation

Chaiken's most noted work previous to *The L Word* was the writing of the Golden Globe–winning telemovie *Dirty Pictures* (2000), which was concerned with the story of Cincinnati Contemporary Arts Center director Dennis Barrie's (James Woods) curation of a Robert Mapplethorpe exhibition and the consequent obscenity trial prompted by the mobilization of the religious right. The character of Bette is clearly drawn from Chaiken's depiction of Barrie: Bette too works as the director of an art museum whose acronym is also the CAC (in this case the California Arts Center), and she spends the latter part of the first season embroiled in rhetorical and literal attacks on the gallery's showing of the "Provocations" exhibition by a right-wing religious group known as the "Coalition for Concerned Citizens" (CCC), led by Fae Buckley (Helen Shaver).

The L Word's "Provocations" exhibition, however, goes one step further in its "provocation" than the Mapplethorpe exhibition, for as well as sexual and sadomasochistic images, "Provocations" includes a fictional video-recorded performance art piece entitled "Jesus is in me" by fictional artist Isabella Pernao. This piece, initially shown as the opening scene of "Luck, Next Time" (1.9),[21] depicts the fictional artist on her knees, actively being sodomized by a representation of Jesus Christ in a classically posed tableau. When the video is rescreened later in Bette's office, she comments on her interpretation of the piece: "Look at her face—she's searching for a feeling. It's like she's longing for faith. She'd do anything to get it. To feel it." A man delivering flowers then asks if the piece is "supposed to be art" and, in response to Bette's affirmative answer, comments that "I guess *Hustler* and *Penthouse* may put out some pretty good art too." The muffled laughter of James, Bette's secretary, affirms that this is indeed a plausible interpretation, though the next shot of the piece, which shows a shot of the artist crouched, still rocking and crying, holding her face in her hands, underlines the complexity of the image beyond that which would be expected in pornographic forums.

Tavia Nyong'o asserts that "[t]he oppositional irony of this piece . . . divides its audience into those who experience it as sacrilege and those, like Bette, who are thrown by it into aesthetic rapture" (104). To limit the piece to merely a divisory "provocation" intended to divide its audience does not, however, truly capture the complexity of this artwork and its inclusion in the series. "Jesus is in me" is suggestive of various interpretations. It could be a literalist interpretation of the sublimated sexual imagery often present in religious discourse and art, punning on the oft-repeated phrase "Jesus is in me," with Pernao's crying thereafter highlighting the juxtaposition of such imagery with religious guilt over sex. Pernao's deployment of the *image* of Christ for her own ends in this artwork can also be metaphorically read in the context of this story line in *The L Word* and its critique of American religious right-wing groups as an accusation that groups like Fae Buckley's likewise use the image and words of Christ for their own purposes. This interpretation is further suggested by Fae's distortion of Bette's words in a video made and broadcast by her group, which both renders the audience disinclined to be sympathetic to her cause and is further suggestive of misinterpretation and reconstitution of *language*, possibly including that found in the Bible, for the achievement of one's own political ends.

Whereas "Luck, Next Time" leaves the question of what is art and what is pornography open while calling upon the viewer to empathize with those who believe in freedom of expression, the next episode, "Liberally," takes a clear position on the differentials between art and pornography. The scene that opens this episode is a segment from the shooting of a pornographic film featuring two school-age "schoolgirls" engaging in "lesbian" activity before one of them is called upon to give oral sex to their male "principal." It is later discovered that this young actress is Cora Buckley, Fae Buckley's daughter, who ran away from home after her mother "couldn't, or wouldn't" stop her father from "abusing the hell out of" Cora (Oscar in 1.10). It is also revealed that Fae had paid off a judge to "expunge the record." This story line both acts to establish Fae Buckley as a morally reprehensible character and to demonstrate her moralistic concern "for the children" to be dubious. It further rhetorically resituates pedophilia, a main argument by the religious right against gay parenting, into the Christian, heterosexual nuclear family, a family form that such groups as the CCC seek to establish as the only possible form of family.

"Luck, Next Time" and "Liberally" are clearly structurally linked by both their opening and closing segments, and these linkages serve to highlight their shared thematic concerns and to strengthen the argument about art and pornography in which these episodes engage. It is significant that the endings of each of these episodes are overlaid with a religiously themed song (Nick Cave's "Into My Arms" and Rufus Wainwright's cover of Leonard Cohen's "Hallelujah," respectively) whose lyrics inflect and affect the audience's response. "Into My Arms" plays shortly after Bette has learned of Tina's miscarriage and immediately following her attempt to comfort her in a darkened house where she is confronted by members of the CCC (including the flower delivery man), who are hammering a sign into her lawn and hurling various epithets at Bette, including telling her that she is "going to hell" (1.9). The song then starts with the words "I don't believe in an interventionist god" and continues as Bette enters the house and rests her head against the illuminated window with arms spread and hands resting on the door frame, silently sobbing, with her framing and posture suggestive of crucifixion. She then straightens up and composes herself, walking off toward Tina and directly into the camera (and thus figuratively into the audience) to the words "direct you into my arms." Despite any feelings the audience may have toward the potentially pornographic nature of the exhibition itself, here the viewer is called upon to empathize with Bette and thereby if not disagree with the perspective of the CCC, at least censure their methods.

"Liberally" ends with "Hallelujah" playing over the intercutting of Bette's tears in the aftermath of her debate with Fae Buckley, scenes of Cora Buckley in the previous video, and new scenes of a blank-eyed Cora soliciting by the side of the road. Fae's assertion to Bette during a television debate that "[t]he Bible condemns homosexuality. That's why God took your unborn child from your lesbian lover . . . and that was a blessing . . . So that he doesn't have to suffer the degradation he would have been subject to had he been born into your depraved life" (1.10) leads Bette to call her a monster and then start crying. The cuts to Cora, who is in both a literally and metaphorically "dark place," evoke Fae's calls upon Bette to feel shame for "making this world a darker place for your child to live in" (1.9), presenting the real dark world created for Fae's daughter by Fae's desire to project and maintain a certain (and false) public image. While Nyong'o feels that Bette's tears in this scene are symptomatic of a "cultural logic of victimization, which constructs the show as a theatre of ressentiment where strength cannot appear except in the hands of a villain like right-wing doyenne Faye Buckley" (105), I feel that this "strength" is portrayed as dubious, and the scenes of Bette crying, together with the scenes of Cora, while they indeed portray victimization, are rhetorically effective in signifying the effects and costs of the triumphs of the religious right and, with the repeated proclamation of "Hallelujah," in questioning what is really being celebrated in the victories of such pieces of "God's work" in which fundamentalist Christians like Fae engage.

The explicit discussion of the demarcation between art and pornography in such story lines as the "Provocations" exhibit in *The L Word* is of further interest in light of claims that *The L Word* itself functions as a pornographic spectacle. Perhaps the most curious example thereof occurred almost exactly fifteen years after (and in the same state as) the opening of the Mapplethorpe exhibition portrayed in *Dirty Pictures*, when a police officer working as a school resource officer was placed on "administrative leave" and possibly fired for having a *The L Word* screensaver. Although the still photographs used in the screensaver "may be far from scandalous," according to the news article, "[s]chool officials in Camden, Ohio, say it's not a problem to have still photographs on a screen saver, but if the photographs advertise a lesbian-oriented TV show, the situation gets complicated" because children may thereby be exposed to "adult issues" ("Sexy Screensaver"). The images themselves are not being argued to be pornographic; rather, the promotional photographs of actresses *playing* lesbians, together with the potentially quite obscure title, are themselves deemed extremely unsuitable for the eyes of children, rendering the very *idea* of lesbianism pornographic or at least fitting the "adult" category.

The L Word has also been critiqued as a pornographic spectacle from quite a different quarter: as "soft porn" that is "appealing to straight men" (Reeder 52) or, as Winnie McCroy put it, "pud fodder for Joe Sixpack." Though others comment that while "there is certainly enough lesbian action on the show to offend the straitlaced . . . there is hardly enough to satiate aficionados of dyke porn" (Nyong'o 104), considering the potential for heterosexual male voyeurism and appropriation of *The L Word*, and particularly in light of Chaiken and Showtime's apparent willingness to market the program to heterosexual male viewers, one can certainly understand Reeder and McCroy's concerns.[22] In the second season of the series, a story line emerges that appears to act as a response to such critiques.

The introduction of the characters Mark and Gomey (Sam Easton) in the second season acts to make intra-diegetic the straight male voyeur that such critiques situate in *The L Word's* perceived audience. A direct-to-video filmmaker whose filmography includes *Wild-Ass Catholic School Girls*, Mark appeals to the women's knowledge of "what it's like to try to figure out how to be an artist" (as well as their financial needs) and tells them that all he really "has ever wanted to do" is "make documentaries" (2.4) in order to gain a position as their housemate. We soon discover that the subjects of his first documentary are to be Shane and Jenny; also, in addition to barely consensual interview footage, Mark places secret video cameras throughout the house and, for the next six episodes, proceeds to film every moment of their lives, at times orchestrating scenarios to make the plot of this "documentary" more interesting.

Initially, the presence of Mark as a voyeuristic villain appears to be a narratively sinister critique of male objectification of women and lesbians in particular. This thematic criticism seems particularly surprising given that *The L Word* has—in its interviews, its advertising, and throughout its various pedagogic apologist moments within character dialogues—attempted to court straight male viewers. Through this story line, the series once again sets out to clearly demarcate the boundaries of "pornography" and "objectification" that will allow the series to remain outside these boundaries. Thus *The L Word*, a product whose very existence appears to be dependent upon the freeing up of presenting sexualized images of lesbians by the proponents of lesbian pornography during the sex wars, interestingly appears to position itself with their (middle ground) opponents: those activists who sought to maintain clear distinctions between "erotica" and "pornography" (see, for example, Steinem's "Erotica vs. Pornography") rather than those who wished to reclaim pornographic images. At the same time, *The L Word* goes beyond what earlier producers of both lesbian erotica and pornography (who sought same-sex-desiring women as their audience) would perhaps have deemed appropriate by explicitly marketing it to a heterosexual (and particularly heterosexual male) audience.[23]

It soon becomes apparent that Mark's presence in the narrative is intended to provide an identificatory viewing position for the presumed straight male viewers of the series. Sitting and conversing with Jenny during his first episode, Mark takes a peek at Shane, Alice, and others in the pool through the same fence that Jenny in the first episode had watched Shane and an unnamed woman having sex, which marked the beginning of Jenny's exploration of her sexuality, prefiguring that Mark, like Jenny, will become fascinated and engrossed by these women in a more than sexual-voyeuristic manner. The contrast made between Mark and his friend Gomey, whose interest in Mark's new housemates is depicted as more prurient than Mark and is reined in by him, is furthered throughout the course of this story line, perhaps most clearly presented when Mark zooms into Shane's face during a videotaped sex encounter because he "just . . . I wanna know what she's feeling," to the protests of Gomey, who asks, "who gives a fuck what she's *feeling*?" (2.6). This somewhat dubious distinction between sexual and emotional voyeurism positions Mark as a narratively more positive and redeemable character than the two-dimensional Gomey. Mark's attempts to capture the essence of Shane's magnetism and to discover the roots of her dissociative melancholy form a kind of worship, suggestive of his documentary being indeed the "art," though exploitative, that he intends it to be rather than the simple pornography that Gomey, and Mark's funding producer, demand. The correlation between Mark's work and that of the producers of *The L Word* is also suggested by the frequent intercutting throughout these episodes between the usual clear, glossy images and Mark's grainy footage, the latter frequently accompanied by shots of him watching the drama unfold. Obviously, this highlights the constructed nature of these very images seen in *The L Word* and the potential for exploitation and objectification that they hold.

Perhaps the compromises the producers of *The L Word* deem necessary to get funding for the series are expressed through Mark's interactions with his funding producer, and the absolution of Mark in the name of "art" and his desire to seek emotional truth is a form of self-explanation for providing the "pud fodder for Joe Sixpack" about which McCroy writes while simultaneously distancing themselves from its more prurient aspects. It is insinuated that Mark or the producers of *The L Word* may be filming the sex, but their wish is to get to the emotional truth of the matter—and they have to show the sex not only in order to do so but to acquire the funding to show people the interior realities of a group of lesbians and thereby to make them relate empathically to lesbian lives. In an interview in the *New York Times*, Alison Glock recounts Chaiken's belief that "'[j]ust telling these stories is in itself a radical act,'" and "if Showtime expects her to tweak and repackage and make lesbianism hot, hot, hot then she is happy to comply" (Chaiken qtd. in Glock 38). Whether one labels this attitude as "pragmatic or opportunist," as Sedgwick puts it (B10), it is clear that some parallels can be drawn between Mark and the producers of *The L Word* in this respect, which is perhaps why Mark is drawn in a somewhat sympathetic light. This story line thereby places both parties in clear juxtaposition to those that *The L Word* unequivocally labels "pornographers."

However, despite the redemptive qualities ascribed to him, Mark's actions are still seen as having a concrete effect of harm, at least on Jenny (though they arguably assist Shane to find happiness with Carmen). Jenny's discovery of the filming appears to trigger a downward spiral that culminates in the surfacing of her repressed memory of being raped as a child, which had been subtly suggested in her writing all season. Jenny makes clear in several forceful speeches to Mark the extent of the violation that his cameras have enacted. The "art" defense the series appears to have given Mark is seen in his own defense of himself when he begs Jenny, "watch my documentary. You know me. You'll understand. It's not what you think it is. I know that I crossed the line. I know that I went too far with this" (2.10). This request is made in spite of Jenny having just told him "don't you dare tell me this is for the sake of art." Jenny responds to Mark's request by asking him if he has sisters, exhorting him to ask "your sisters about the very first time that they were intruded upon by some man, or a boy. . . . Because there isn't a single girl or woman in this world that hasn't been intruded upon and sometimes it's relatively benign, and sometimes it's *so* fucking painful." (2.10)

Violation is herein not excused in the name of art, and the production of this kind of surveilled pornography is placed in *The L Word* on a continuum of sexual violence, seen in the reference to Mark's "rapey cameras" (2.10). Despite this and various other speeches that consider Mark and his actions unforgivable, Mark not only remains on the series until the end of the season, but despite his actions, a certain degree of narrative and audience sympathy for him has been built up over the course of his arc, and by the final episode his penitence seems to have been rewarded by his being offered an opportunity to narratively "go legit" as a filmmaker by filming a speech by Gloria Steinem. How Ms. Steinem, a key figure in the discourse against pornography and female objectification, might feel about such a person filming her is not discussed. His remaining in the series is the result of Jenny's daring him to "stay here and deal with this," specifying that "we're not friends" (2.11). This acts as a potential dare to the heterosexual male viewer consuming the series voyeuristically to not be shamed into not watching the show as a result of this story line but instead to learn to watch it in a different way and atone for his voyeurism via the feminist consciousness implied by Mark's taping of Steinem.[24]

Jenny has also, earlier in the season, acted as the mouthpiece for much of the discussion of the objectification of women and lesbians. Hunter, a young man in her writing class, reads out a Henry Miller–style story, which Jenny alleges turns "all your women [characters] . . . into these nameless, faceless, body part whores" (2.7). To Hunter's claims that "*That is not true. I—I honor my woman characters. I imbue them with sexual agency,*" Jenny replies,

> Your main character, Jasmine, she, like, opens up Madeleine's world by giving her the best fucking orgasm she's ever had—*[Hunter looks a little taken aback]* which, I don't know if you know this, is the primary sex act that two women can actually have! And then *you* go ahead, and you belittle it by turning it into pornography, and I think that the reason why you're doing this is because men can't handle it, the fact that these women can have this *amazing, fucking, beautiful, mind-blowing orgasm, without a fucking cock!* (2.7)

As well as acting to further establish lesbian sex as "real fucking," an ongoing project of season two, the scene clearly establishes a rejoinder to those who seek, by literary or personal means, to appropriate lesbian sexuality into a pornographic framework for heterosexual male consumption.

The ambiguous attitudes the series presents toward sadomasochism and deliberate pornographic display are then realized during the aftermath of the Mark story line. Jenny, still reeling from Mark's invasion, visits an S/M dungeon known as "The Seven Stations of the Cross" (2.11). We have earlier seen Dana and Alice, who are currently engaging in sexual experimentation with strap-ons and gendered role-playing, enter the venue in curiosity and leave immediately, somewhat frightened. Jenny later ventures there after another confrontation with Mark, less in the name of sexual experimentation than in the enactment of a kind of grief. Immediately after being tied to the cross, she fully recovers her submerged memory, and the audience finally discovers what happened to Jenny and how her present self still inhabits this originary scene of violation. Her almost-engagement in masochism has been successful in regaining a memory that she has attempted previously, by storytelling and a makeshift prayer ritual, to recover so that it will not continue to haunt her. In contrast to these devices, the cross works because, it is suggested by *The L Word*, a masochistic act is a subconscious reenactment of her trauma, as seen when she looks frightened and disassociative and cries upon being tied down.

Through an arrangement with the dominatrix, in "L'Chaim" (2.12) Jenny begins to work at a very seedy strip bar, asserting that her choice to do so is motivated by a desire to control her objectification:

> Because, when I'm in there, it's my fucking choice when I take off my top and I wanna show my breasts. And it's my fucking choice when I take off my pants and I show my pussy, and then I stop when I wanna stop and it makes me feel good because I'm in charge, and it helps me remember all this childhood shit that happened to me. You know, like, I have to. It's important. (2.13)

The earlier part of this statement is reminiscent of lesbian stripper and cofounder of lesbian pornographic magazine *On Our Backs* Debi Sundahl's assertion that

> [w]omen who work in the sex industry are not responsible for, nor do they in any way perpetuate, the sexual oppression of women. In fact, to any enlightened observer, our very existence provides a distinction and a choice as to when a woman should be treated like a sex object and when she should not be. At the theatre, yes; on the street, no. (qtd. in Dolan 63)

While Sundahl's and—it would initially appear here—Jenny's declarations act as a reclamation of stripping as at least not harmful, if not potentially empowering, in Jenny's case this activity is juxtaposed with a scene later in "Lacuna," where she is discovered by Shane on the floor, crying and covered in blood as the result of many acts of self-injury. This, coupled with her memories that form the genesis of both of these activities, is suggestive of her stripping being less a controlled "choice" than a form of masochism, which is represented as analogous to self-harming behavior. David Calof argues that "self injury expresses unresolved trauma and disowned affect" and can be used to "re-enact previous abuse as a way of feeling a sense of control that was not present during the original abuse" (qtd. in Vivekananda 21–22), an understanding that is consistent with Jenny's stripping activities as well as with her cutting. *The L Word* thereby presents masochism as being performed by those not of sound mind, sexual exhibitionists as necessarily victims, and all these quite different acts within a kind of continuum of violence against women that includes rape and incest.

Upon close examination of *The L Word*, it does not appear to have the celebratory or even accepting attitude toward pornography that would perhaps be expected from such a mainstreamed, post–sex wars product that is clearly willing to present and market itself as a form of soft porn. Instead, a clear distinction is postulated between "pornography" and "art" reminiscent of Steinem's distinctions between erotica and pornography, with the former being understood as "freely chosen, mutual sexuality," while the latter is seen as existing within a continuum of violence against women (240). This dichotomy at times appears to be rather amorphous and self-serving; pornography is situated as anything *outside* that which the characters or metaphorically the producers of *The L Word* engage in, while the most violating practice, Mark's tapes, is to a degree deemed "art" through their focus on the emotional state of the characters and metaphoric linking to the producers of the series, although their nonconsensual nature is chastised. The constant counterposing of these distinctions allows *The L Word* to simultaneously condemn those who would seek to censor potentially pornographic art and perform a narrative condemnation of pornography. Although in each of these circumstances the pornography depicted is made by and intended for straight males, the lack of reference to the genre of lesbian-authored and intended pornography makes it difficult to ascertain *The L Word*'s attitudes toward the latter type of pornography.

CONCLUSIONS

The L Word outdid its record second season network pickup by having a third season announced before the second season even began to be screened. Whether this pickup would have occurred had Showtime known that the second season was to achieve much lower ratings than *The L Word*'s opening season—it is reported that the premiere of the first episode of season two gained 50 percent less audience members than the series premiere at a little over half a million ("*The L Word*: Season 2 and 3 News")—is unknowable, though the real viewership, if one takes into consideration Showtime's multiple rescreenings, international free-to-air and cable broadcasting, sales of DVD box sets, and internet tape trading and downloads, is significantly higher. Due to a "ratings explosion in season three," Showtime renewed the show (Greenblatt qtd. in Showtime Networks Inc.), and the sixth season, due to be screened in 2009, is to be the final season of the show. With its immense international popularity and six season run, *The L Word* will always hold a seminal place in the history of lesbian pop-cultural representation. Due to this cultural significance, the relatively high numbers of viewers for a lesbian cultural product, and the influence its advent has had upon the discourse surrounding lesbianism in the early years of the new millennium, it is vital to attempt to analyze *The L Word*'s attitudes toward and representations of lesbian culture and to ascertain the discourses that it relies upon and purports to assess whether it really represents what it appears to on cursory examination.

The presence of Steinem as a special guest star in "Lacuna" functions perhaps as a gesture from a fairly postfeminist series to the presence of feminism in lesbian history and an assertion of the continued need for feminist activism in contemporary society. Unfortunately, the scene in which *The L Word* characters converse with Steinem is rather trite and suffers from very poor dialogue that renders the whole scene rather confusing if not absurd. All that is really established via this conversation is a rather preachy message that there are feminists who are not lesbians; that there are lesbians who are not feminists; that straight women are bigger "man-haters" than lesbians; and that some are "predisposed"—to *what* exactly it is not immediately clear due to the scripting, but it seems that this conversation refers to being predisposed to (any) particular sexuality—while some have a choice regarding which gender they are attracted to. Steinem then reappears to give a speech that acts as a liberal-feminist call to action. The inclusion within *The L Word* of such a call in some respects acts to cycle it back to an earlier understanding of what would be included in a lesbian text in a manner that is initially somewhat surprising, but upon deeper examination of *The L Word*'s positioning in regard to the battles fought during the sex wars, this is not really so.

Despite its much lauded or criticized new femme visibility that the series appears to offer, upon deeper perusal, its relationship with lesbian femininity is complicated and at times rather troubled. In *The L Word*, it does not appear to be the case that "it is the femmes who are finally asserting themselves" (199), as Clark speculated of late eighties and early nineties lesbian chic, for femmes and indeed lesbian femininity are not necessarily authenticated by the series, which instead undertakes a strange practice of simultaneously making feminine lesbians hypervisible and rendering them less authentic. The series' attitudes toward other products reclaimed by the lesbian sex wars, such as butch-femme relationships, nonmonogamy, pornography, and sadomasochism, appears equally narratively combative. The journey of lesbians into mainstream consciousness enacted by *The L Word* thus represents an odd combination of the usage of the depoliticized, willing to be objectified images (made possible during the sex wars), while simultaneously disavowing the practices and identities that allowed and encouraged these images. The series does seem to perform something unique, and I think this is situated in the sheer multiplicity of the clichés activated in *The L Word*. This multiplicity creates an encyclopedia of lesbian cultural representation on the one hand, and on the other it offers a certain productive space for reflection on new forms of representing lesbians on television.

CHAPTER 5

Dressing Up, Strapping On, and Stripping Off

Contemporary Lesbian Pornographic Cultural Production

WHILE *QUEER AS FOLK* PORTRAYS ITS LESBIAN CHARACTERS AS THE "sex-negative" protagonists in a discourse on sex and assimilation and *The L Word* seeks to posit clear distinctions between art and pornography and further portrays pornography as part of a continuum of violence against women, these popular cultural products temporally coexist with lesbian products and cultures that continue the celebratory attitudes toward lesbian sexual cultures and lesbian pornography popularized during the sex wars.

There is some disagreement over who won the sex wars. Many writers engaging with the issue would contend that the shifting cultural terrain of the eighties and nineties indicates the victory of the sex radicals, but others are not so convinced. While Emma Healey asserts that "[l]ove it or hate it, by the end of the [1980s], a new language of sexuality and a new fashion for sexuality had been created and the voice of one political ideology had been challenged and almost silenced" (6), Maureen Engel sees the Butler decision in Canada as indicative that antipornography forces have gained institutional power: "[i]f 'The Sex Radicals Won the Sex Wars,' as my fridge magnet contends, it is a hollow victory when the Anti-Pornographers are winning over state apparatuses" (115).[1] While Kathleen Martindale sees the sex wars as having "set the terms and the agenda for contemporary feminist and lesbian discourses on sexuality and sexual representation" (7), Tamsin Wilton argues that "[t]he sex wars, it appears, have not been won or lost; they have simply become unimportant" (2). The continued presence of the central issues and methodologies of the sex wars in the contemporary settings I have identified so far in this monograph would seem to point to just such a lack of a firm conclusion to these debates, and the continued importance of sex wars issues in contemporary lesbian culture and cultural products.

It has been argued that the blossoming of lesbian pornographic cultural production that ensued during and due to the sex wars was era-specific and that this era has now passed. Engel, concluding her doctoral thesis "Arousing Possibilities: The Cultural Work of Lesbian Pornography" proclaims,

> When I began this project, I had only an intuitive idea that something in the world of lesbian sex-and-porn wasn't quite the same anymore. What I've come to realize over the course of this writing is that while this project began as a contemporary one, it is now an historical one; the moment that is both the subject and object of this study has passed, or has at least been immeasurably altered. (213)

Through the identification and analysis of various forms of contemporary lesbian pornographic cultural production, this chapter contends that "the moment" of "lesbian sex-and-porn" is far from over and indeed continues into the present day.

THEORIZING LESBIAN PORNOGRAPHY

Reviewing the "wave" of American lesbian sex magazines for *Gay Community News* in 1984, Judith Butler remarks,

> So many of the fantasies given in the pages of these journals are efforts to enter into another world, a world where struggles between women are confined to the erotic, where difference is dynamic and engaging. The anonymous woman who claims or is claimed sexually by another, the perfect understanding, the orgasm that lasts for hours or covers whole continents— this dream of a release from pain which can only happen by embodying pain seems pervasive and real. Sexual puritanism, whether enforced by dominant ideologies or radical feminism, seems bent on thwarting sexual self-expression, even where that expression is crucial for attaining feminist ideals of autonomy and freedom. ("Desire and Dread" 3)

Within this statement, Butler emphasizes the political importance of lesbian sexual self-expression, seeing such expression as essential to attaining autonomy and freedom. This insistence on sexual self-expression as *political* is a central feature of much lesbian sex-radical discourse and indeed lesbian pornography. Not all manifestations thereof, however, interpret this political dimension within a framework of "feminist ideals." Butler concludes her review by asking,

> So what has happened to the radical feminist movement that *On Our Backs* threatens to displace *Off Our Backs* as the journal of lesbian culture? Think about it: *Off Our Backs* is a call for resistance to the patriarchy. *On Our Backs* is a call to desire. The language of anger has been supplanted by the need for fulfilment. Clearly, there is a place for both, and with an enhanced dialog in the pages of these journals, we might well understand how to speak both languages at once. (7)

Butler's recognition of the concurrent need for resistance and desire and of the advantageous nature of understanding "how to speak both languages at once" is an important one. Her continued discussion of feminism in this article sees these new sex-radical publications in dialogue with feminism(s). Indeed, the themes that Butler sees emerging in these journals—of reclaiming sexual pleasure, reconceptualizing women's bodies, and recognizing "how inner struggles become part of our erotic struggles with other women" (3)—are easily identifiable as subjects frequently at issue in feminist consciousness-raising groups and on the pages of lesbian-feminist publications. Yet this recognition of lesbian sexual publications as being part of a continuing or evolving feminist dialogue is frequently sorely missing from later academic and popular discussion of these sex-radical cultural productions and cultures.

In Terralee Bensinger's 1992 essay "Lesbian Pornography: The Re/Making of (a) Community," for example, "feminism" is a universalized, hegemonic, global force that acts to repress differential, subjugated, local "lesbian" sexual expression. Citing a major cause of lesbians' problems as being "caught between the moral imperatives of the feminist 'law' and their personal, private desires which may exist in violation of that law" (70), Bensinger takes issue with the location of "an enemy" outside the feminist movement (i.e., the patriarchy): "By targeting an 'enemy' that exists 'outside' the movement, conflicts existing inside, at the level of everyday practice, are obscured" (69). While this indeed may be the case, Bensinger's apparent contention that the lesbian and feminist movements[2] should turn their fighting energies inward and that the notion of an "outside enemy" is rather foolish seems hazardous if not downright dangerous. The example that seems to be given to justify such beliefs in Bensinger's article is that of the lavender menace, whose "strategic attempt to counteract homophobia . . . located all lesbians squarely with the category Woman, an identification that was politically and not erotically based" (74) and thus produced a "fiction of a unified community" (69), laying the foundational bricks of many binarisms. An example given of such binarism is "feminist/lesbian," though how exactly the lesbian-feminists of the lavender menace, known throughout the article as either cultural feminists or radical feminists, managed to create a binary *between* lesbian and feminist through declaring that the two are enmeshed is not made clear by Bensinger. Bensinger reads this action as "an attempt on the part of politically identified lesbians to gain acceptance through complete assimilation into the feminist movement" (74).

Mary T. Conway's "Inhabiting the Phallus: Reading *Safe Is Desire*" likewise is still invested in reading lesbian pornography in the context of "the anti-penetration position of U.S. cultural feminism," though it also addresses it in terms of "the Lacanian theory of subjectivity," in order to analyze the manner in which the film "not only deconstructs, [but] asserts a new natural" (135, 133). The tendency to oversimplify "the feminist" and overburden her with negativity located by Wilton, as well as Biddy Martin, Dana Heller, Teresa de Lauretis, Bonnie Zimmerman, Sue-Ellen Case, Shane Phelan, and Suzanna Danuta Walters (as is discussed in the introduction to this monograph), is also apparent in "Inhabiting the Phallus," although unlike such writers as Bensinger who generalize an amorphous feminism, Conway at times uses the more specific term "anti-penetration cultural feminism" (134) as the field her subject and analysis is arguing against. In Conway's essay, however, "anti-penetration" is used both as a signifier of itself and as a signifier of antidildoism. The inextricability of her usage is telling, and the conflation points toward an understanding of the two as literally identical, which has the function of *both* furthering the myth that lesbians who do not use dildos do not engage in penetrative sex *and* encouraging the conflation of feminist and lesbian critiques of dildo use with the more extreme perspectives that critique all forms of penetration. This is despite Conway's own critique that cultural feminism's denial of "the exercise of phallic power . . . relied, of course upon a conflation of phallus/penis, a manoeuvre simple in its execution, but complex in its practical discursive effects" (143).

Of course, such oversimplification of feminism and feminist involvements in the sex wars is not universal to the critical discourse. In Lisa Henderson's "Lesbian Pornography: Cultural Transgression and Sexual Demystification," for example, Henderson describes herself as part of the "feminist anti-anti-porn contingent" (174). Wilton's *Finger-Licking Good: The Ins and Outs of Lesbian Sex* also takes a more measured approach to the subject, acknowledging that "[m]any of the most important defenders of lesbian sexuality, many of the most productive makers of lesbian erotica and/or representations of lesbian sex, are feminists" (10), while still being critical of the "lesbian sex prefects" (10) and the negative effects "the feminist deconstruction of male supremacy and patriarchal heterosex" had in its labeling of "more and more behaviours ... as male supremacist power play" (11). That is, despite being critical of many aspects of feminist theory and practice, Wilton acknowledges the importance of "thirty years of feminist activism" to the evolution of contemporary lesbian culture and queer theory, making it clear, in contradistinction to writers such as Bensinger, that "feminism is *not* the enemy" (10) and critiquing those whom she sees as "trash[ing] lesbian-feminism unreservedly" (125). Emma Healey in *Lesbian Sex Wars* even goes so far as to assert that the sex wars "were the wrong wars with the wrong targets" and that "it is wrong to use sex, how we think about it and how we practice it, as the measure of what a good or trendy lesbian should be" (14).

A cultural or theoretical movement's ability to perform such measured analysis is perhaps the surest sign of growth and development. A further example of development is the emergence of texts that look beyond the original subject of debate. The materialization of a theoretical discourse analyzing the lesbian pornographic and other cultural products of the sex wars is a promising indication of further sophistication of the field. This may also assist in combating the ephemerality of the genre and lessen its marginality and disreputable status in the academy. Engel notes the following about early critique:

> As I began to read what had been written about pornography, lesbian and heterosexual alike, I was struck by the sheer absence of the materials themselves; reading pornography seemed to involve reading an idea of pornography—creating a representation of representations that provided a modicum of safety, and perhaps even a veneer of respectability. (2–3)

Engel seeks to rectify these problems via a direct analysis of the magazines *On Our Backs* and *Bad Attitude*, the work of the "performance/art collective" Kiss & Tell, and the writings of Joan Nestle and Pat Califa.

She is not, however, alone in her meticulous attentions to the texts themselves. Henderson's "Lesbian Pornography" discusses *On Our Backs* and Califa's *Macho Sluts*, and Wilton, in her monograph *Finger-Licking Good*, scrutinizes the magazine *Quim*; the photography of Della Grace, Tee Corinne, Laurence Jaugey-Paget, and Kiss & Tell; and the work of "lesbian sex gurus" Califa, Nestle, JoAnne Loulan, and Susie Bright, as well as providing a history of and engaging with the theoretical debates of the sex wars. Conway has published two consecutive articles on Blush/lesbian pornographic films, while almost all analyses of the theories and histories of the sex wars contain at least some direct engagement with lesbian pornographic products.

In "What Do You Call a Lesbian with Long Fingers? The Development of Lesbian and Dyke Pornography," published in the anthology *Porn Studies*, Heather Butler provides a thorough history and analysis of "cinematic lesbian sex acts and actual pornographic films from 1968 to 2000" (167). Butler identifies several key periods of lesbian video pornography, including that of lesbian sexploitation films of the late 1960s and 1970s; 1980s lesbian pornography such as *Erotic in Nature* (1985) which "is extremely indicative of what has been happening (politically, socially, culturally) among lesbians in the 1970s and 1980s" and "constitutes one of the first attempts to create a lesbian presence in pornography as *different*; that is, as distinctly nonheterosexual in its emphasis on the more erotic aspects of lesbian sexuality" (175). Butler also discusses early 1990s "dyke" pornography that "is safe-sex savvy and not afraid to appropriate sex acts once considered definitive of heterosexual and gay male pornography, such as penetration, dirty talk, rough sex, and role-playing" (181–82). Later 1990s pornography, which includes "more radical changes, among them younger and more urban bodies, extensive dildo- and role-play, nonmonogamous sex, dirty talk, and a continuation of the safe-sex education themes which emerged in the early 1990s" (185) as well as amateur dyke pornography are also discussed by Butler.

LESBIAN PORNOGRAPHIC CULTURAL PRODUCTION POST-2000

This chapter in essence takes up where Heather Butler left off in its focus on lesbian pornographic cultural production after the millennium. Having introduced the academic discourses with which I am in dialogue, I will now introduce the contemporary cultural products that have motivated the inclusion of this chapter. While these by no means form an exhaustive list of texts available within the genre, each highlights particular issues or showcases trends that resonate with the issues of mainstreaming, gendering, spectatorship, and sexual practices that have been under discussion in other texts throughout this monograph. The contemporary pornographic cultural productions I will particularly focus on in this chapter include the films *Sugar High Glitter City* (2001), produced by San Francisco–based S.I.R. Video; *Tick Tock* (2001) and *Madam and Eve* (2003), both produced by the U.K.-based company Rusty Films and directed by Angie Dowling; and *The Crash Pad* (2005), the first pornographic film from new company Pink and White Productions. I will also examine lesbian striptease and burlesque in the form of Sydney troupe/club *Gurlesque*. Several of the features Conway elucidates in regard to some specific Blush/Fatale lesbian pornographic video texts are also of significance in these later texts: "the education of the lesbian porn spectator" ("Spectatorship in Lesbian Porn" 103), questions of authenticity and the display of "a diversity of body types" (95), and the reevaluation of the gaze. These factors further resonate with the history of lesbian pornographic production, particularly that which has emerged since the early 1980s.

Sugar High Glitter City takes place in the fictional Glitter City (shot in San Francisco), where sugar is outlawed. This premise lends itself to scenarios involving prostitution for various forms of sugar and includes such characters as police officers, a sugar madam, a reformed sugar addict, and multiple young women who prostitute themselves for sugar. Openly classifying itself as within "the dykesploitation genre" (DVD back cover), this film is very much in the realm of what Heather Butler characterizes as "dyke porn," with its appropriation of the conventions of heterosexual and gay male pornography, extensive use of dildos, rough sex, role-play, and an urban setting (181–82, 185).

In contrast, *Tick Tock*, a U.K. film directed by Dowling ("Rusty Cave"), while it too features dildo scenes, appears more reminiscent of earlier models of lesbian pornography in its "emphasis on the more erotic aspects of lesbian sexuality" (Butler, "What Do You Call a Lesbian" 175). One extended scene, for example, simply shows a woman running her hand over her lover's body and then doing the same with kisses. Also directed by Dowling and featuring as its cast "members of a real-life sex commune" (Rowlson), *Madam and Eve* is set in a "hospital" named Eden whose purpose is "sexual awakenings." This latter film is more highly stylized; features more dildos and more elaborate, fetish-inspired costuming; and has higher production values than its predecessor. Comedy plays an important role in the interludes between sexual scenes, from the opening segments wherein Madam gazes into her crystal ball, to a wall of moving eyes under the sign "ICU" (Eve tells a new nurse who is taken aback by them, "oh, don't mind them, they just like to watch"). This, together with the title, the name of the hospital, and the "treatment" of "The Promised One," gives *Madam and Eve* a very campy tone, which is further reinforced by kitsch sets, music, and acting style.

The most recent film under discussion, *The Crash Pad* (2005), is directed by San Franciscan director Shine Louise Houston, who has been prominent in the promotion of the film, including interviews in such publications as the U.S. lesbian magazine *Curve* (Lo, "She's a Very Dirty Girl" 42–43). The ostensible narrative revolves around "The Crash Pad," an apartment to which lesbian couples go to engage in sexual intercourse. One must be given the "key" to the apartment and various rules apply, including only using the apartment seven times before handing the key on to another. Houston's university training in film and her experience with several short experimental films renders her pornography more thoughtful than other examples of the genre. For example, her films include complex explorations of identity and metatextual gestures that will be analyzed later in this chapter. When asked in an interview given as an extra on the DVD edition about to her "angle" as a pornographic filmmaker, Houston stated "I'm sure as you can see from this cast I definitely wanna get some more diversity.... I think women-made porn just has a slightly different edge to it ... [and] being a person of colour, that brings a little bit [of a] different angle.... [T]here's a big gap it seems like for women-made dyke porn." ("Director's Interview")

During cast interviews, it becomes clear that several of the actresses are real-life couples, and this—together with Houston's assertion that "pretty much anyone can come up to me and say 'hey, I wanna be in your porn'" and that she expects to gain new participants from those who watch the video ("Director's Interview")—emphasizes both a certain authenticity of depicted lesbian desire and a potential community focus in terms of viewers becoming participants.

Such fluidity between performers and consumers is also a key feature of *Gurlesque*, as will be discussed later in this chapter. *Gurlesque* is a women-only lesbian strip/burlesque club, and the name is evocative of both the club itself and a troupe of regular performers. *Gurlesque*'s performances began at the Imperial Hotel in Erskineville (Sydney), Australia, in September 2000, and they have since regularly performed at several other venues, including in Melbourne, Brisbane, and Adelaide, before being relocated to the Marquee Club in Camperdown (Sydney) during late 2005, and now appearing in the well-known Sydney nightclub Arq.[3] Performances take place approximately twice a month during the *Gurlesque* season, which lasts from October to April/May to coincide with Australia's warmer months. Structurally, performances are very similar to drag shows in that they take the form of several hourly shows, each including multiple acts framed by MCs and organizers Sex Intents and Glita Supernova. The acts themselves comprise many different types of performances, from "straight" strip to severely confronting performance art, stand-up comedy routines to camp parodies, drag kinging to high femme performance, explicit political statements to artful elegies. The linkage between these acts is that clothes are generally taken off or, occasionally, put on. Each act is given full creative autonomy; the only guidance given to the performers is to "explore the essence of striptease" (Intents and Supernova).[4]

SPECTATORSHIP

Lesbian and dyke pornography generally positions itself toward the lesbian spectator, particularly through its attempts to distinguish itself from straight-produced "all girl" pornography that purports to display "lesbians" and is directed toward a heterosexual male spectatorship. Some lesbian pornographic texts perform this by claiming lesbian authenticity extradiegetically, as seen in S.I.R. video's "seal" sporting the guarantee "100% dyke produced" on its DVD covers or in Rusty Films' insistence on the covers of *Madam and Eve* and *Tick Tock* that they are "100% lesbian made," a claim given further credence by having Red Hot Diva, a subsidiary of the U.K. lesbian magazine *Diva*, as their codistributors. What about their context, however, within the adult entertainment industry, an industry largely catering to male spectators? Despite their (often politically motivated) efforts to challenge or create lesbian sexual cultures, these production companies exist and survive within a very particular heteronormative and male-centric economic system. Looking from a purely economic standpoint, producers of video productions such as these, like the producers of *The L Word*, may indeed wish to attract, or at the very least unintentionally succeed in attracting, a more general audience of male voyeurs attracted to "lesbian" pornography. Indeed, *Madam and Eve*'s all-femme cast, noted even on the back of the DVD as "[a]n all-stunning latex-clad femme cast of 7 luscious lesbians," could be argued, like *The L Word*, to use the femininity and conventional beauty of its performers to market this commercial product to both lesbian and heterosexual male audiences alike. Such productions can thus be viewed as participants in commodifying lesbian sexuality in a manner that other smaller-scale cultural productions, such as *Gurlesque*'s women-only stripshows or the Australian independent lesbian sex magazine *Slit*, can avoid.

Tick Tock goes further than the other texts under consideration in its claims to lesbian authenticity by proclaiming on its cover that it features "real lesbians enjoying real sex." While this could be read as an attempt to prove to a lesbian audience that it is not an example of male-centric faux "lesbian" pornography, it could also be read as potentially appealing to an ethnographically curious gaze, a desire by males to "know" lesbian sex. The latter fits into the traditional mode of hard-core pornography that "obsessively seeks knowledge, through a voyeuristic record of confessional, involuntary paroxysm, of the 'thing' itself" (Williams 49). Should the latter, however, indeed be the case, such viewers would seem to gain little extra knowledge from *Tick Tock*, which frustrates viewerly expectations in its inattention to the conventions of pornography. (Whether this is in deliberate appeal to a different form of pornographic spectatorship or due to inexperience on the part of the filmmaker is unclear.) The goal-oriented nature of much pornography is structurally undermined in *Tick Tock* by cuts, at times in the middle of a sequence, to another couple, then back again, in a style more reminiscent of television narrative that wishes to portray its action as happening simultaneously than of other pornography, where the journey to orgasm is usually visually depicted in as much detail as possible.

These cuts provoke a disjointedness that is at odds with goal-oriented pornography and decentralizes the importance of orgasm, a decentralization that is, as Heather Butler has noted, a feature of other examples of lesbian pornography (187). Indeed, orgasms, or rather portrayals thereof, are infrequent in this video, with some scenes left still in process at the conclusion of *Tick Tock*. The "very difficulty involved in the representation of lesbian sex acts" (177) that Butler identifies through *Erotic in Nature* is also highlighted in this video, where at times one can see little but the back of a cast member or a tangle of limbs shot in close-up, with cuts to the faces and back again, leaving the viewer at times quite perplexed as to what is actually happening between the couple. The cuts away from the "action," together with the periodic difficulties in distinguishing exactly what the "action" is, frustrate viewer expectations and leave a certain shroud of mystery around "lesbian" sex. Cutting between various lesbian couples, however, also emphasizes notions of community in terms of its evocation of the idea that multiple lesbian sex acts are going on at any given time and offers its spectators the possibility of participating in a community of lesbian sex through their engagement with the film itself.

Tick Tock's claims as to the "realness" of its sex scenes are somewhat undermined by the presence of a clearly staged "story line"; nonetheless, there are features that do highlight a certain lesbian authenticity, if not necessarily the enjoyment that is likewise touted on the cover. First, the performers do not appear to be professionals; this is ascertained not only by the nonmainstream ideals of beauty they represent but also by the decidedly amateurish nature of the film. The performers frequently burst into giggles in the midst of their scenes, supporting legs slip and the performers fall awkwardly, and there is communication between the performers followed by frequent changes in position when something does not "work." Such aspects, while positing a certain degree of not performing *lesbianism* (much of the ostensible narrative is indeed portrayed via cartoon-inspired "thought bubbles"), do, however, at least in some cases, appear to undermine the "realness" of the performers' enjoyment and to resituate the role of the spectator. Instead of being offered a narrative that requires a certain suspension of disbelief but adheres to filmic conventions that situates the pleasure of the spectator at center stage, viewers of *Tick Tock* are exposed to a pornographic film that, while clearly made with a lesbian viewer in mind (one performer even has posters of Ellen DeGeneres and Martina Navratilova behind her as if to "prove" her lesbian credentials), does not appear to be explicitly focused on pleasing its viewership but rather is more invested in the *creation* of lesbian pornography itself.

Theorists of lesbian sexual representation have long grappled with questions of spectatorship and address. This focus is often posited as a response to "[f]eminist film and performance critics [who] argue that representation is addressed to the gaze of the male spectator" (Dolan 59). Looking through the perspective of a psychoanalytic discourse, which locks them into a dichotomy of men as bearer of the active "look," critics such as Laura Mulvey see little way out of women's entrapment as objects short of "trans-sex identification" (Mulvey 33). Jill Dolan notes that "[p]art of the problem with the psychoanalytic model of spectatorship is just this tendency to pose universal 'male' and 'female' spectators who respond only according to gender" (61). Lesbian critics have sought to problematize this tendency by introducing the lesbian gaze into the realm of debate—a gaze that both situates women as potential active bearers of "the look," and dissolves "[t]he subject/object relations that trap women performers and spectators as commodities in a heterosexual context" (Dolan 63).

Dolan's "Desire Cloaked in a Trenchcoat" engages specifically with a theoretical examination of subject/object and economic relations within various strip clubs, first examining the positionality of female performers in heterosexual contexts and then posing the "women-only strip show" (63)[5] as allowing its performers to "assume the subject position [through being butch and dressing in masculine attire] rather than objectifying themselves. The butch/femme role play allow[s] the performers to seduce each other and the lesbian spectators through the constant of lesbian sexuality" (63). Moreover, "[t]he recognition of mutual subjectivity allows the gaze to be shared in a direct way," creating "a new economy of desire" in which "[r]ather than gazing through the representational window at their commodification as women, lesbians are generating and buying their own desire on a different representational economy" (64). However, despite her critique thereof, Dolan's analysis still works within the psychoanalytic context she critiques by framing the assumption of the subject position as requiring butchness: masculine (though female) identification. This approach is emblematic of much theorizing of lesbian subject positions, discussions that are generally unable to account for the femme as a gazing or a powerful subject. Dolan does, however, acknowledge in an epilogue that "[t]here is no universal lesbian spectator to whom each lesbian representation will provide the embodiment of the same lesbian desire. Sexuality, and desire, and lesbian subjects are more complicated than that" (65).

Bensinger and other writers on lesbian pornography have also cited its potential to transcend "the male gaze" through displacement; the simple removal of the male spectator from the viewing realm renders the problematic aspects of "the male gaze" inconsequential. Although such a perspective is in some respects valid, it is also to a degree an oversimplification because it presents a universalistic model of lesbian desire and spectatorship (as Dolan notes) and represents "the male gaze" in such monolithic and powerful terms. Perhaps power relations in specifically erotic spectatorship are more complicated than initially thought in discourses influenced by a highly rigidified gendering of power and desire. Katherine Liepe-Levinson's work "Striptease: Desire, Mimetic Jeopardy, and Performing Spectators" is based on three years of field research on American heterosexual strip events.[6] The article views the conventions of these various events primarily in terms of spectatorship and where the conventions of striptease break down or uphold conventional sex roles, a process that "interrogates several of the prevailing theories about sex, gender, representation and the dynamics of desire" (10). Liepe-Levinson concludes,

> The Dworkin-McKinnon, Marxist-materialist, and psychoanalytic critiques of sexual difference and social inequality have revealed much about the roots and structures of sexism and other kinds of bigotry in our culture. However, these theories are problematic because they rest on two-dimensional formulations of desire, which the strip show, being a performance specifically about desire, foregrounds. (32–33)

Such "theories of gender and representation cannot thoroughly interrogate how sociosexual symbols are used in the construction of female and male sex roles because their formulations eliminate so many of those uses," further denying sexual agency and "reinvent[ing] and promot[ing] the double standard that insists that explicit sexual representations are inherently harmful to women" (33). Liepe-Levinson's research suggests the necessity for a more complex understanding of heterosexually oriented sexual performances in terms of their power relations and discursive effects. As such, perhaps a more complex theorization of the potential and varied significations of lesbian sexual performance is necessary than the simple declaration of it as powerful due to the absence of the male gaze in order to avoid the universalism Dolan rightly acknowledges.

Other writers question the very division between the active desiring gaze and the passive (narcissistic) identification with the objectified woman. Reina Lewis and Katrina Rolley, for example, discuss the consumption of lesbian-coded images in mainstream fashion magazines by both lesbian and heterosexual women readers. They suggest that "the magazines themselves almost invite—indeed educate—the [heterosexual female] reader into something very close to a lesbian response in much of their imagery of women in general," implicating them in "a paradigmatically lesbian viewing position in which women are induced to exercise a gaze that desires the represented woman, not just one that identifies with them" (181). Lewis and Rolley locate lesbian spectators as being positioned to simultaneously desire the woman in the fashion photograph and want to be that same woman being (presumably) desired by other lesbians. Having identified these dual capacities, Lewis and Rolley assert,

> This raises the question of whether/how far one can distinguish an overtly lesbian gaze, that self-consciously desires the represented woman, from a narcissistic one that identifies with the represented woman as an object of a

presumed-to-be-male desire? . . . either desiring for herself the represented woman or/and desiring to be the woman who is desired by other lesbian readers. (181)

Such readings complicate and potentially undermine the conception of split gazes (masculine-active-objectifying/feminine-passive-narcissistic) in lesbian visual pleasure by maintaining that the gaze can be (or perhaps necessarily is) both one of distancing (understood as objectification) and narcissism: "the double movement of a lesbian visual pleasure wherein the viewer wants both to be and *have* the object" (Lewis and Rolley 181).

What of the contemporary lesbian spectator? Lesbian cultures have certainly gone through periods where spectatorship of pornography was considered a politically problematic internalization of misogyny and male identification. During the sex wars, with the cultural blossoming of lesbian-produced pornography and sex radical cultures, participating in and viewing lesbian porn was often viewed as a transgressively political gesture. Where are contemporary lesbian cultures positioned in terms of these histories? While the production of lesbian pornography has, as I have articulated above, continued into the present day and indeed has branched out into new fields such as comics,[7] growing numbers of, and support for, politically and socially conservative platforms—the "new homonormativity," as theorist Lisa Duggan puts it—would seem to engender a culture more reluctant to embrace dyke porn.

Pondering these issues at the beginning of 2004, I attended a significant number of the lesbian-themed films screened during the Sydney Gay and Lesbian Mardi Gras Film Festival. While the majority of the lesbian films I attended screened to almost empty theatres, two films screened as part of the festival played to packed houses: notably *Mango Kiss*, billed as a "good old romantic comedy of errors . . . only this time of the role-playing, nonmonogamous, S/M dyke variety," and *Madam and Eve*, billed as "lesbian cinema's sexiest and most visually exciting erotic film ever" (Queerscreen QS-13).[8] The overwhelming popularity of an S/M comedy and an erotic film would seem to suggest the cultural currency of sex radical culture—yet audience reception to the latter, more sexually confronting film was certainly not uniformly positive.[9] While this can be explained by either the film's own deficiencies or its promotion, which to some extent obscured the actual content of the film (it would be more honest to characterize it as pornography than as an "erotic film," as there was little to no plot), it would not seem to account for the extremity of the audience reaction. Audience members started leaving the theatre almost immediately, and halfway into the film, there were less than half the initial audience numbers remaining. Many audience members who did remain either spent their time laughing or looking extremely uncomfortable, and they continued to gradually leave throughout the screening. This was, however, not the first time the Mardi Gras Film Festival had shown lesbian pornography as part of its program. In 2001 they showed S.I.R. video's *Hard Love and How to Fuck in High Heels*, with *Sugar High Glitter City* being screened the following year. Had audience response to the previous two films been overwhelmingly negative, surely the festival organizers would not have chosen to once again program a lesbian pornographic film.

Perhaps this reaction simply points toward the necessity of the education of the lesbian porn viewer. Conway has discussed how a lesbian porn film can "educate . . . its intended emergent lesbian viewer on another subject: how to be a lesbian porn viewer. . . . re-educat[ing] one sort of female viewer (previously absent, perverse, or punished), transforming her into a desiring lesbian spectator" ("Spectatorship in Lesbian Porn" 100). This is often achieved by bringing lesbian porn spectators into the frame of lesbian (video) pornography, who then "serve as models for the education of lesbian produced porn viewing" by performing "spectatorship and voyeurism" (102). In the context of the Blush/Fatale film *Safe Is Desire*, for example, the inclusion of intradiegetic spectators complicates fixed economies of desire because "[t]he audience members model subject-status, but also are depicted as objects, for each other and for the extradiegetic spectator . . . [which] enables the dizzying, simultaneous occupation of both subject and object for nearly everyone in the video" (106–7).

The first sequence of *The Crash Pad* likewise uses an intradiegetic voyeur. As in Conway's discussion of *Safe Is Desire* in which the main characters "serve as models for the education of lesbian produced porn viewing" (101), the presence of the voyeuristic Simone offers lesbian viewers unused to the consumption of pornography a desiring, voyeuristic subject on which to model themselves. Unlike in *Safe Is Desire*, however, where the economies of desire are rather free flowing, Simone is not positioned as an object of desire. Not only is she fully clothed, shots of her are apart from the pornographic "action" and generally take the form of close-ups of her face in an eyeline match with the scene that is taking place, prompting the viewer to follow her gaze rather than gaze upon her. It is significant that one of the participants in this scene (Jiz Lee) had initially come into "The Crash Pad" with Simone to have sex and attempts to leave when noting that the venue is already in use. Simone stops Jiz, prompting her to engage in sexual intercourse with the two prior inhabitants for Simone's viewing pleasure. This narrative logic promotes the pleasurable possibilities of sexual voyeurism and is further suggestive of sex-radical ideologies of nonmonogamy. A flashback point of view sequence in black and white from Simone's perspective toward the end of the sequence further highlights her positioning with the viewer. This sequence shows Jiz kissing Dylan after her orgasm. That it is this gesture of intimacy that is recalled by Simone in a point of view flashback and that it is then suggested as pleasurable for Simone via a close-up on her face further offers to its viewers a cultural logic that discourages possessiveness in sexual partnerings.

A further level of intradiegetic voyeurism takes the form of the metafilmic inclusion of the director and the revelation at the ending that the crash pad itself is a "set-up," filled with hidden cameras by the director. *Tick Tock* also features a character who observes the other participants via hidden CCTV cameras, with frequent shots of the character masturbating to her screen—providing a model, like Simone, for the lesbian porn viewer. *The Crash Pad*, however, takes this trope further by highlighting the role of the filmmaker, generally absent from pornography and, indeed, from film more generally. Early in *The Crash Pad*, we see the director (Houston) give the key of "the crash pad" to another woman, outlining the "rules" of the venue. During this scene, Houston is shot from a high angle, over the shoulder of the character to whom she is giving the key, thus positioning the viewer and this character as being given the metaphoric "key" to the "crash pad" via the film itself. During the final sexual sequence, a solo masturbation scene, the subject ("Jo," who had previously been involved in the first threesome scene that Simone observed) gazes fixedly into the camera, with frequent cuts away to black and white footage and shots from different angles. After Jo's orgasm, she looks directly at the camera and then laughs, turning away. The scene ends with a focus on Jo's face, zooming out in color as Jo gives a little wave to the camera/viewer. A wipe of a water-ripple during the zoom-out transforms the color image into a black and white image, which is revealed to be on a computer desktop with another shot of the scene, obviously part of an editing process that, it is then revealed, is being presided over by Houston, who ends the film by turning to the camera, smiling, and raising her eyebrow playfully before turning back to her work.

This segment not only reveals the narrative of hidden cameras and a deliberate filmic voyeurism but also positions the viewer as complicit in

Houston's secret watching of these women's sexual lives. This scene also, through Jo, demonstrates an exhibitionist willing to be watched, which acts to ethically authorize this device and the viewer's voyeurism. This, together with the presumed though not sole presence of a female audience, renders this story line narratively quite different from Mark's secret videoing of Jenny and Shane in season two of *The L Word*. It is also quite unlike the positioning of the female voyeur in *Tick Tock*, who is "caught" at the end by one of the women she has been observing and is playfully told that she has been "[b]usted. You know I could call the police and put you in jail for this. But you'd like it too much." The—albeit playful—specter of police in *Tick Tock* is nowhere present in *The Crash Pad*, as Houston is authorized not only by Jo's knowingness but also by her coy interrogation of the act of filmmaking itself as an act of voyeurism.

During live lesbian pornographic events such as *Gurlesque*'s lesbian strip shows, there is an added dimension to the deliberate drawing of the spectator into the frame undertaken in these films, as there is no extradiegetic spectator. All spectators are subject to themselves being objectified both by other members of the audience and by the performers themselves. The often elaborate costuming of many *Gurlesque* audience members further positions audience members as performers attempting to elicit the gaze of other spectators. That audience members indeed perform at *Gurlesque* and that performers frequently join the audience to view one another's acts further undermines distinctions between performance and spectatorship within the venue. The realities of *Gurlesque* also render more complicated theorization that posits the simple removal of the male gaze as automatically changing power dynamics in the realm of the stripshow and ensuring the subjectivity and indeed safety of performers therein. This was perhaps most powerfully demonstrated by the very first *Gurlesque* performance. Recounting the event, Supernova notes that at the first *Gurlesque* the female audience "didn't know how to behave, they were grabbing at the strippers, and Sex had to go out and give them a lecture and tell them how it is, how you need to act. . . . [I]t was worse than working at the Cross[10] that night" (Intents and Supernova). Contrasting this to the generalized docility of the (male) audiences Liepe-Levinson documents in her work on heterosexual stripshows and indeed to current (female) *Gurlesque* audiences indicates that the power relations of the stripshow do not exist solely upon a gendered axis and articulates instead the centrality of spectator education in attaining and retaining the subject power of performers in a sex performance environment.

REPRESENTING LESBIAN GENDER

Another important feature of proving lesbian authenticity within lesbian pornography comes in the form of the character Conway describes as "the most visibly deviant lesbian, the most serious contender for the male gaze, and as such the character absent from 'All-Girl' mainstream porn: the butch dyke" ("Spectatorship in Lesbian Porn" 106). Such theorists of lesbian pornography as Heather Butler see the butch as the figure who "authenticates lesbian pornography, even if only superficially. . . . Without being convinced of 'real' pleasure or a real orgasm, the viewer may nevertheless acknowledge a lesbian authenticity through the figure of the butch" (169). The display of butch actresses, or those who otherwise do not fit within mainstream ideals of female beauty, within lesbian pornography can further be seen as bringing a specifically lesbian erotic economy to productions as well as being indicative of a certain political intent that wishes to renegotiate accepted standards of which bodies are to be deemed attractive or erotic.

Sugar High Glitter City features a multiplicity of visual styles and body types, from coproducer and performer Jackie Strano's portrayal of a butch undercover detective, to a scene involving a drag king and a middle-aged boi engaging in sexual activity visually identifiable with gay male sex, to the high femme sugar "Madam," who has a fuller figure than actresses generally seen in conventional pornography. This "diversity of body types" (Conway, "Spectatorship in Lesbian Porn" 95) is often a feature of lesbian pornography and acts to authenticate the film and differentiate it from "lesbian" pornography, in which the actresses are generally feminine, large-breasted, and, quite disturbingly, often have long fingernails. Beyond such authenticating aspects, these inclusions bring into play questions of gendered representation and gendered power dynamics.

Madam and Eve, conversely, does not display any butch actresses. Following Conway and Butler's perspectives, some may argue that the femininity of *Madam and Eve*'s actresses renders it less authentic and more invested in catering to the male gaze. Indeed, with its fantastical scenario and elaborate latex costumes, including the fetishlike nurse outfits, *Madam and Eve* could be said to be little different from heterosexual pornography. If we agree with Dolan's discussion of (live) sex performers in a lesbian environment claiming subjectivity through performing the butch role discussed earlier, these performers would be viewed as necessarily entrapped, due to their femininity, as objects in the realm of desire and spectatorship. But is feminine subjectivity, or at least agency, truly impossible within such a context? Within most examples of lesbian pornography that I have investigated, it is indeed the butch(er) one who is almost always the sexual top, furthering the ideological connections between masculinity and active sexuality. When a masculine lesbian is topped, it is almost always by another masculine lesbian, which does not undermine this characterization. Within *Madam and Eve*, however, perhaps due to this absence, there is room for the depiction of an active feminine sexual agency, most clearly articulated in the character of Eve.

The character of Eve is markedly feminine, yet through her dominance and sexual prowess she displays an active feminine sexual agency. This occurs most distinctively in the scene reviewers most often note in relation to *Madam and Eve*, which depicts Eve having sex with two women simultaneously, utilizing two dildos strapped to her thighs to pleasure the women. While entering into the phallic visual economy through the use of the dildos, the repositioning and doubling recontextualizes this so that within this sequence the feminine woman not only is depicted as the sexual top but also is rendered as exceeding and thus potentially undermining the schema of supervirility usually associated with masculinity.

Unlike in *Madam and Eve*, in *The Crash Pad* a multiplicity of gendered styles are seen. This includes an obviously butch-femme couple (who, it is revealed in an included interview, are a couple in real life) as well as a threesome involving a feminine woman and two more masculine women with a feminine intradiegetic spectator. Each of these, however, does share the conventionalized schema of the butch(er) participant as necessarily the sexual top. The complication or difference of this from heterosexual pornography is, however, that it is indeed the feminine, seemingly "passive" character upon whose pleasure the activity is focused. Thus, she is not portrayed as a passive object as in much heterosexual pornography, but rather she is an active, desiring subject. Indeed, in the first sexual sequence of *The Crash Pad*, the feminine Dylan directs the actions of the two more masculine women who are ostensibly topping her, regularly switching positions to optimize her own pleasure. That the intradiegetic voyeur is also on the feminine spectrum implies and encourages the viewerly identification with the one being penetrated, which encourages a different logic of spectatorship from much heterosexual pornography wherein the viewer is encouraged to identify with the (male) sexually "active" participant. The conventions are also subverted in that the orgasmic culmination of the scene takes place via Dylan's digitally stimulating her own clitoris, while her two partners lie on each side of her, stimulating her nipples, rendering the "active" sexual partner, if not superfluous, then at least foreplay to the "main dish" of an empowered female sexuality that knows what it wants.

A later sexual sequence in *The Crash Pad* depicts a boi-boi couple. As Felice Newman observes, "Two bois enter the pad and proceed to wrestle for dominance—this is a visual reference to a standard in the canon of gay male pornography in which the loser's butt is the trophy of the victor." Despite such referents and a certain level of adherence to these conventions, there is shortly some switching between the partners. The conventions of gay male and heterosexual pornography and the power relations depicted therein are further undermined by the female ejaculation scene. Heather Butler proposes,

> Lacan's formula, and the subsequent feminist critique of its formulation, implies that the phallus belongs to the man, yet the lesbian with her object of penetration ["whether it comes in the form of a dildo, long fingers, a tongue, or any other object" (184)] can perform all the same things that the penis/phallus can perform during the sex act, except for one very important thing—she, or rather her dildo, does not ejaculate. . . . This stands in direct contrast to the end of a typical sex scene in heterosexual pornography, in which a man will ejaculate on some part of the female body. . . . [F]emale protagonists in heterosexual pornography . . . turn, for the most part, into cum-catchers, and there is little attempt to represent female pleasure in any form other than a smiling or ecstatic face dripping with semen. (184–85)

Butler is using this discussion to argue that "a strap-on dildo provides the kind of agency to a woman (or two women) that a man's penis simply does not" (185), noting but not positioning as central that "her partner (. . . may, in fact, ejaculate)" (184).

Toward the end of this sexual sequence in *The Crash Pad*, the "loser" of the initial wrestling match (Jiz Lee), who has been the primary bottom in this sequence, after much penetration, ejaculates copious amounts of fluid, to which her partner (Shawn) responds by positioning herself underneath Jiz and "drinking" it. This is clearly a visual reference to the convention of "[e]jaculation on the woman's face" or "facial" (184) that Heather Butler discusses, yet it also undermines this convention through the reversal of power and sexual agency it performs. This scene in *The Crash Pad* repositions visually depictable erotic agency within the female body, subverting these conventions of ejaculation rather than simply removing them. This sexual sequence, which does not contain any dildos, empowers female sexuality and recontextualizes "passive" female sexuality beyond the simple removal of the male or the male gaze by rendering ejaculation the role of the penetrated partner, realigning notions of active and passive sexuality.

In Linda Williams's *Hard Core: Power, Pleasure, and the "Frenzy of the Visible,"* Williams cites the work of Getrud Koch, who asserts that "all film pornography is a "drive for knowledge" that takes place through a voyeurism structured as a cognitive urge. Invoking Foucault, Koch argues that film pornography can be viewed as an important mechanism in the wholesale restructuring of the experience of sexuality into a visual form." (48)

The principle of maximum visibility can be seen in the manner in which "the genre has consistently maintained certain clinical-documentary qualities at the expense of other forms of realism or artistry that might actually be more arousing" (48). With the principle of maximum visibility guiding the genre, difficulty arises in the restructuring of the experience of female sexuality into a visual form, for "while it is possible, in a certain limited and reductive way, to 'represent' the physical pleasure of the male by showing erection and ejaculation, this maximum visibility proves elusive in the parallel confession of female sexual pleasure" (49). Williams goes on to note that "[h]ard core desires assurance that it is witnessing not the voluntary performance of feminine pleasure, but its involuntary confession" (50).

The aforementioned scene from *The Crash Pad* makes it possible to "represent" the physical pleasure of the female by visibly depicting a "confession" of sexual pleasure in much the same manner as it has been for males in the genre of hard-core pornography. While the sheer volume of fluid that Jiz Lee ejaculates may make the viewer question whether or not this indeed has been "faked" in some way, the volume is also suggestive of a form of involuntary "confession." If the filmmakers had wished to fake such a thing, they would undoubtedly have taken a more modest approach to the quantity of ejaculate in order to convince their viewers of the "veracity" of the scene because authenticity is clearly important in this video, as evidenced by the included interviews that are at pains to exposit that the women enjoyed the sex acts they engaged in for the making of the film. The scene also takes on the "educational" model Engel sees as a feature of such print forms of lesbian pornography as *On Our*

Backs (80–87). As J. Lowndes Sevely and J.W. Bennett observe, "Culture and language tend to obscure knowledge that the human female has a prostate gland and is capable of ejaculation" (1). The question of whether women do indeed ejaculate, which Sevely and Bennett back in 1978 felt could not yet be answered conclusively, has since been determined by sex researchers through its being *made visible* to them via the anecdotes and testing of women who experience it (see Belzer, Addiego et al. and Heath), and indeed made further visible by such products as the Fanny Fatale (Debi Sundahl, cofounder of *On Our Backs*) sex educational video *How to Female Ejaculate* (1992). There is still, however, a great deal of ignorance about the possibility of female ejaculation, and visible depiction such as that seen in *The Crash Pad* demonstrates to women via visual representation the sexual possibilities of the female body of which they may previously have been unaware.[11]

In the case of other, less clearly visible forms of orgasm, Houston encounters the same problematics of visible representation of female sexual pleasure that is discussed by Koch and Williams. In the director's commentary over Jo's sexual sequence, Houston comments,

> The question of the female pop shot comes up again, so how do you show, *visibly* show something that's hidden, not exactly visible, especially with somebody who's not like, super, like, verbal or . . . doesn't spasm as much as say, Dylan, like in the first scene. So there's still a lot of questions, a lot of problems, that I have that I want to work out like in future films. . . . I'm reading lots of Linda Williams, a book called *Hard Core*. . . . Its definitely good food for thought. . . . [H]opefully that'll help me make smart, informed pornography in the future, tackle this problem of the female pop shot.[12]

The scene to which Houston here refers is the final masturbation scene with Jo, discussed earlier in regard to the hidden cameras, which acts as a very complex expression of gendered identity. Jo is at first depicted as holding a dildo to herself and stroking it as though it were indeed a symbolic penis, then penetrating herself with the same dildo. After various such switches, she finally orgasms by manually stimulating her clitoris while holding the dildo to her, implying some kind of transgendered or bigendered sexual subjectivity. In the director's interview, Houston notes that one aspect of this scene is "touching the bases of, y'know, my weird dual identity" that includes both masculine and feminine sides. Thus, the representation of Jo as expressive of an element of Houston's own sexually gendered identities, performing to the hidden cameras that Houston herself controls, creates a further doubling effect.

As almost all discussants of lesbian pornography have noted and as can be seen in several of the sequences discussed here, dildos tend to play a fairly significant role in the genre. The issue of what place a dildo, or even a drag king, has in a lesbian stripshow was raised at the Mardi Gras recovery (from the parade's after party) *Gurlesque* show in March 2004.[13] During this show, MC Sex Intents wore a strap-on dildo for the majority of the production, [14] and there were more drag acts and dildos than is usually the case during *Gurlesque* performances. Following an act that featured two zombie drag cowboys sodomizing each other to the tunes of Bon Jovi, Intents asked a member of the audience to repeat to the crowd a sentiment she had expressed to a friend during the act. The statement was along the lines of "what is the point of lesbian sex with dicks" or "what exactly does this have to do with lesbian sex?" The crowd, containing a large number of women from S/M and lesbian pornographic communities, reacted to this with outrage and jeering. Intents's response was in contrast fairly sensitive and intelligently framed. She began by sympathizing with the questioner's feelings before moving to an explanation that involved exhortations for freedom and multiplicity of sexual expression, together with a designification of the dildo as phallus, elucidating the dildo as simply an object to be used for sexual pleasure without further heterosexual inferences.

The latter is clearly reminiscent of Bright's injunction, "Ladies, the discreet, complete and definitive information on dildos is this: penetration is only as heterosexual as kissing. Now that truth can be known! Fucking knows no gender. Not only that, but penises can only be compared to dildos in the sense that they take up space" (19). Bright's statement raises the alluring prospect of resignification, of the redefinition of objects seemingly unrecuperable to lesbians and lesbian sex. Engel further and more fully articulates this sentiment:

> While a society (or in this case, even a subculture) may imbue particular sexual practices with particular meanings, it is important to recall that those meanings do not necessarily inhere in the practices themselves. . . . Bright's mantra is not simply one of liberation, demystifying and throwing off the repression that has surrounded this particular sexual practice; it makes available a different way of seeing that allows readers to recognize not only the meanings that they make, but also the meanings

that they might *un* or *re*make by reinvesting new meanings into sexual practices. (86)

Some not only see the potential to reinvest new meanings into certain sexual practices but also place a great deal of emphasis upon the power and significance of the dildo and its potential to deconstruct, or sometimes even radically alter, power relations via the claiming of the phallus—the ability to signify in Lacanian/psychoanalytic-feminist terms.

Discussing the ramifications of dildo usage in the lesbian porn film *Safe Is Desire*, Conway asserts that "[t]he natural has not merely been deconstructed, but a new natural asserted, represented by gender parody and the inhabitable, siliconic dildo" (155), which is seen to have the effect of "delinking chromosomal sex from behavioural sex" (152). As with notions of the gaze, other theorists have rejected the notion that such Lacanian-influenced analyses are even applicable in this context. Heather Butler, for example, contends that "the very idea of the phallic is displaced when the desire represented is lesbian" (184), proposing that "the Lacanian phallus has as its telos not penetration, but rather ejaculation. This gives a whole new meaning to the word *lack*; for one could argue that there is no lack into lesbian sexuality, that the real lack is in Lac(k)anian psychoanalytic theory" (185). Butler, however, like the other discussants, still situates the dildo as a radical feature of dyke pornography. For her this is obviously not due to its positioning as the phallus—as she points out, "[d]ildos are accessories to the lesbian sex act; they are in no way requisite" (184)—rather, she sees the "conscious commodification of sex toys" in lesbian porn as "effectively displacing the commodification of female bodies that typically occurs in heterosexual porn" (191).

Others are less inclined to view dildos as a neutrally gendered signifier. As Lisa Henderson remarks:,"Unlike Susie Sexpert's perky reassurances about the benign dildo in "Toys for Us," Nestle was moved by how dildos *could* be used in the story to connote subversions of conventional gender identity and the tense (if consensual) play of sexual power between or among women partners." (181)

Such a reading seems to be appropriate in an analysis of the first sexual sequence in *Sugar High Glitter City*. During the first sexual sequence, the use of dildos together with the traditionally gendered power displays of the narrative scenario render the scene visually little different from heterosexual pornography. The scene depicts two "sleazy corrupt detectives fuelled by their own wicked addictions"—Blue (Jackie Strano) and Stark (Stark)—who sexually "shake down" Bitta Honey

(Brooklyn Bloomberg). The dialogue and power dynamics of the scene seem to depict this as skirting the borders of nonconsensual sex.

There is in this scene, however, also what Henderson refers to in her discussion of *On Our Backs*'s erotic fiction as "[t]he tempering effect of demystification" (179). In her reading of these stories, Henderson identifies moments where the protagonist is "rescued from a dubious image of coercion by acknowledging her own desire to be 'handled'" (179–80). This sequence of *Sugar High Glitter City* also attempts such gestures to prevent the image falling too far into politically dubious terrain. This can be seen in the voiceover that begins the scene, which asserts that some "don't want protection" from the detectives, and in the later seeming transformation of the scene into one that is focused on Honey's pleasure, a pleasure that is made visible to the audience via Honey's ejaculation and her vagina forcefully expelling Blue's hand upon orgasm (it is significant that for this latter portion of the scene, they have dispensed with the dildos).

QUEER AND FEMINIST: UNDERMINING THE DISTINCTIONS

The insistence on the political nature of displaying female/lesbian sexuality is also an important feature of much lesbian pornography. Engel, in "Arousing Possibilities: The Cultural Work of Lesbian Pornography," argues that the "timeless spaceless setting" widely noted as a heterosexual pornographic convention is absent in lesbian pornography, as "lesbian sexuality . . . always exists in a political context that cannot ignore history or geography" (20). Unlike many heterosexual pornographic productions, lesbian porn is generally engaged with a sense of itself as a political gesture. *Gurlesque* is certainly a manifestation of this phenomenon, both in its general positioning and in recurring narratives within individual acts involving the throwing off of oppression or repression.

In a 2003 review of *Gurlesque*, Natalya Hughes discusses the show in the context of Walters' critique of queer theory and practice, aptly situating the radical potentialities of *Gurlesque* as being precisely in its positioning at the juncture of queer and feminist politics:

> *Gurlesque* might be best seen as straddling the fields of feminist theory and queer theory insofar as it acknowledges and celebrates the fluid sexuality of queer, but not at the expense of a consideration of the specificity of gender. At its best, *Gurlesque* might be seen to perform the implications of these debates at the level of practice, working towards the visibility of lesbian feminism via particularly entertaining means. (2)

Walters' "laundry list of the queer contemporary" includes the following signifiers: lesbian strippers, "in your face activism," go-go girls, *On Our Backs*, drag, piercing, tattoos, cross-dressing, butch-femme, dildos, and camp ("From Here to Queer" 830). Obviously, in *Gurlesque* these signifiers are of central significance. Indeed, the very existence of this cultural production itself is predicated upon queer culture and theory and the sex-positive lesbian culture engendered by the sex wars. Walters (as I have outlined more fully in the introduction) rightly critiques many aspects of this (then) "new queer culture"—most particularly the hyperbolic enunciations of "queer" that pose themselves in opposition to, or even as, "the antidote to a 'retrograde' feminist theorising" (832). Walters, along with other theorists such as Martin, suggests a reinsertion of gender into the frame of sexuality theory—a lesbianism that is not automatically positioned in opposition to a frequently distorted view of the feminist past in order to correct its perceived mistakes. *Gurlesque* seems to present something that I think Walters would never have expected: a unique synthesis of feminist ideas and queer culture within the setting of a lesbian strip show.

Hughes asserts that to

> dismiss the show as merely oppositional or antagonistic to feminist politics however, would be too simplistic, ignoring some of the most fundamental aspects of the *Gurlesque* phenomenon, which, as I understand it, seemed to convey quite explicit engagements with feminist concerns. As the group's promotional material suggests, *Gurlesque*'s objectives align it with issues that dominated second wave feminism, in its attempts to provide a forum in which to "confront fears and insecurities, to challenge taboos, both sexual, social . . . [and] to explode myths about . . . what a patriarchal society[15] dictates is sexy or attractive" The format of the show echoes this "consciousness raising" objective, as between acts, spectators are encouraged to not just watch, but also take an active role in the night's proceedings. (1)

Hughes's cogent observations are further evidenced by the privileging in *Gurlesque* of identification *with women* over and above identification on an axis of sexuality/queerness, a connection that initially appears anachronistic in light of the perceived shift from lesbian-feminism's identification based on gender to lesbian postmodernism's identification based on sexuality. This can be seen in the creation of a women's-only space (notably, unlike some other such spaces, this includes transgendered women) in the variety of sexualities of both performers and attendees and in the repeated assertions about *women's* bodies and *women's* sexualities. With the exception of Intents and Supernova, who frequently refer to themselves as lesbians, the sexuality of many of the *Gurlesque* performers is indefinable and not asserted, although several of the performers are visually identifiable as lesbians through their profiles or participation in Sydney's lesbian community. This focus is in quite direct contradiction to both perceptions of "the new queer culture" as identifying along lines of sexuality as opposed to gender and conventions of lesbian/dyke pornographic production in that the performers' authenticity as "real lesbians" is not thereby asserted. Both during their onstage discussions and when being interviewed, Intents and Supernova frequently referred to and emphasized the centrality of *women*, rather than solely lesbians or queer women, to their mission:

> [T]he thing is with *Gurlesque* it's like, yeah, it's a lezzo strip joint, but its also a place where *women* can go together, and it doesn't matter if you're bi or gay or straight, married, single, you know, its just about women getting up there together. 'Cause a lot of the performers are, you know, all sorts . . . we wanna support strippers, we wanna support the strip industry, we wanna support the art industry, and you know, women in general to explore. (Supernova qtd. in Intents and Supernova)

Interestingly, *Gurlesque* cofounder Meredith cites two varieties of opposition to initial *Gurlesque* performances: radical feminists *and* queers. The most negative and overt response seems to have been from the latter group: "there's a lot of queer people kinda seem to think that we've put the queer movement back about ten years cause we have a women-only space when we have the club" (Meredith). This led to various *Gurlesque* posters being defaced with slogans like "separatist bullshit." Meredith also observed: "ironically it's a lot of people who are like lesbian radical feminists . . . initially got upset about it, I don't know where they're with it at the moment, but they were very upset about it saying they were objectifying female bodies and that it wasn't very progressive in terms of fighting patriarchy and all that kind of stuff." Intents later noted that these latter disgruntlements did not result in any action as such: a planned picket did not eventuate due to the issue splitting the Sydney University Women's Action Collective "down the middle" (Intents in Intents and Supernova).

Much as radical lesbian-feminists sought to deconstruct hegemonic beauty ideals, the insistence upon "exploding myths about body structures" is a familiar theme in lesbian-produced pornography. In *Bathroom*

Sluts, for example, the actresses cite one of "their motivations for participating in lesbian produced porn" as a hope to "represent a diversity of body types, to combat feelings of unattractiveness experienced in response to the dominant culture's standards of beauty" (Conway, "Spectatorship in Lesbian Porn" 94), and this can also be noted in most of the contemporary lesbian pornographic films discussed in this chapter. Such diversity is certainly at least the intention in *Gurlesque* shows, although it does not always succeed in practice. In *Gurlesque* a further dimension is added via the conscious and frequent emphasis on participation, which is made possible by its "live" status and elucidated especially by verbally calling upon audience members to themselves become performers regardless of body type, either in a one-off strip to earn prizes or to sign up for an act in a later show. One *Gurlesque* show (25 Apr. 2004), in fact, opened with an act in which Intents and Supernova narratively enacted this call. Intents, clad as a clown in a tutu with newspaper-stuffed breasts, was shown auditioning for a stripshow and was told very rudely that her various attributes were insufficient. She then removed her fake breasts and auditioned for *Gurlesque*, reaffirming herself as sexy and comfortable in her body. After this performance, the MCs renewed the call for performers, asserting that women of all shapes and sizes are beautiful and sexy. Although performers who conform to at least some conventional standards of beauty greatly outnumber those who do not at *Gurlesque*, there have been various performers who queer the boundaries of what society deems to be an erotically acceptable body.

Gurlesque also does not uncritically embrace its own medium of pornography. While *Gurlesque*'s endeavors to "provide a space that allows women to explore and interpret the essence of striptease" (*Gurlesque* Web site) could be presumed to simply valorize pornography in the name of sex radicalism, there are performances that would seem to agree with Walters' assertion that "this new (uncritical) embrace of porn seems somewhat empty" (858). During a November 2004 show, Supernova performed an act that functioned as a critique of the pornography industry and of the commodification of women's bodies more generally. Appearing in a gold frock and skullcap headdress, Supernova was flanked by gold cutouts of a large dollar sign and a television screen (emblazoned with the logo "FONY"). At the end of the first song, the purpose of these became clear as she shifted from dancing to performing as a newsreader within the borders of the television screen. A voice-over parody of a news broadcast began with various bawdy jokes before shifting to a more serious tone. In this "special financial report," it was announced that wages for strippers had hit an all time low. The act continued on to discuss various aspects of commodification. Crossing her arms over her breasts, Supernova-as-newsreader asserted that it was illegal for women to go topless in public in order to position female breasts as rare commodities, "so they can sell our bodies." Supernova then returned to dancing, and gradually stripping, to the tune of Billie Holiday's "Love for Sale." During this scene, Supernova displayed a vacant, inviting stare, and her movements were more mechanical than in her usual acts to highlight the performative element and to diminish the audience's feelings of intimacy with her. Once again returning to the television screen, Supernova discussed the sanitization and standardizing of women's bodies in the name of pornography and the ownership of women's bodies by the porn industry. After performing to two more songs—"Money Money" from the musical *Cabaret* (Ebb et al.) and Nina Simone's "I Want a Little Sugar in My Bowl"—and after stripping down to complete nakedness, Supernova ended the act by putting black stockings on her feet and hands and then a black bag over her head. With only her naked torso visible, she shook her breasts, commenting upon the exploitation and dehumanization of women's bodies in pornography.

Performances such as this act to further contextualize what has been seen by Walters and others as the setting up of a new hierarchy in which "[s]trippers, hookers, and other sex workers were once pitied for the abuse they received at the hands of the patriarchy, now they are applauded as the heroines for a sex radical future" (858). While a section of the lesbian community may applaud them, strippers and other sex workers still exist and work in a society where they are significantly oppressed and dehumanized. The significance of sex radicalism is that it seeks to refigure women from their roles as hapless victims of the sex industry to powerful "sex-positive" lesbians to empower them and to emphasize the centrality of sex to lesbianism. As is seen in Supernova's performance, there are limitations to such an approach. However, the challenge to hegemonic culture here issued is not censorship of pornography, but rather a strip show that situates one of the causes of such exploitation as the false production of rarity and thus a commodity, thereby positing *more* public display of bodies as a solution. It also functions as an attempt to make the audience conscious of the deeper implications of their own erotic gaze, a gaze that is encouraged, and in fact demanded, by *Gurlesque*. It is not the erotic gaze itself that is here questioned, for the erotic gaze is celebrated in *Gurlesque*, but the manner and levels of respect in the gaze. When asked about this act, Supernova was quick to point out, "I'm pro sex-industry, but I'm seeing this side that . . . we're so owned. . . . [I]t's not just the sex industry, it's the advertising industry. . . . [W]e're so marketed as women that we're not free" and to assert, "I think there should be room for women to critique the industry if they've been in it," but "you can't judge it unless you've been there" (in Intents and Supernova).

The iconic Joan Nestle, who in fact spoke at the 1982 Barnard conference, which is frequently cited as sparking the sex wars, also writes of attending a touring *Gurlesque* production as the conclusion to the new introduction of *A Restricted Country*. Referring to *Gurlesque* as a "feminist lesbian strip ensemble," she writes,

> [F]our women strutted across the stage, their pasties in the shape of missiles, their faces covered by masks of John Howard (the Australian Prime Minister), George Bush, Tony Blair, and Saddam Hussein. Each sported expandable cocks of different sizes—Howard's was the smallest. When they turned their backs to the audience, they all revealed another mask, the screaming-face portrait by Edvard Munch, his human elemental cry of pain. The women twisted and contorted their bodies, thrusting their cocks at each other and taking different positions of dominance. Hussein buggered Bush, Bush fucked all the others. The dance closed with the women waving two placards in the air, "oil" and "blood." I looked around me and thought, [t]his is just where I want to be—in this most restrictive time, when my country's government is in the hands of scoundrels, when capitalist fundamentalists are building their empire, let the beautiful bodies of young queer women mock their power. Let us never give up the struggle to win back our international humanity, let us not despair at the seeming victory of the heartless. Into the streets with all we know of pleasure and the joy of difference. (xiii–xiv)

With this last sentence, Nestle echoes her oft-quoted sentiment that "[b]eing a sexual people is our gift to the world" (xvii). While the celebration of pleasure and camp-mockery are cited as central elements, through her exposition of this particular act she also identifies the other political engagements of *Gurlesque*. Unlike *Queer as Folk*, for example, which, as was seen in Chapter 3, positions queer sex as pure radicalism without heed of other forms of oppression queers may experience, such *Gurlesque* acts as this appear to view "homophobia [as] inextricably linked to sexism and racism and militarism and classism and imperialism. And a few other things" (Bechdel, *Dykes and Sundry Other Carbon-Based Life-Forms* 2) in a manner more reminiscent of the multi-issue politics of days gone by than the fractured single-issue politics favored by the larger contemporary gay and lesbian political organizations.

While, as Engel notes, some earlier sex-positive cultural products "name[d] themselves as not only outside of, but actively and consciously antagonistic towards, Radical Feminist politics" (70) or ridiculed "wimmin love and sisterhood" (Lumby 34), *Gurlesque* appears to have grown beyond such antagonism, instead synthesizing feminist and queer cultures and thereby demonstrating through practice the definite relations between two ideologies theoretically considered to be widely divergent. Even a cursory viewing of contemporary lesbian pornographic production reveals attitudes and positionalities that are widely divergent from the characterization of lesbians seen in such televisual texts as *The L Word* or *Queer as Folk*. In different ways, each of these productions continues a particular ideological trajectory of lesbian culture that came to prominence during the sex wars. Their presence not only acts to acknowledge the histories of lesbian pornographic production but also builds upon these legacies, revealing the diversity of contemporary lesbian cultural representation and production.

Chapter 6

Dykes to Watch Out For and the Lesbian Landscape

WHILE *GURLESQUE* SYNTHESISES ELEMENTS OF RADICAL FEMINIST AND propornography ideologies, the comic strip *Dykes to Watch Out For* (1983–) provides a cultural product in which multiple, at times competing, ideologies are presented in tandem, allowing for not only the expression of various perspectives but also the recognition that such diversity does and can coexist, contrary to what fashion-conscious epitaphs of each lesbian generation would purport. *Dykes to Watch Out For* is a lesbian-centered comic strip, authored by Vermont-based Alison Bechdel, which has evolved from a single-panel comic drawn in 1982 and first published in 1983, to a strip format addressing various lesbian stories and issues in 1984, to a steady cast of characters in 1987 whose lesbian lives, politics, and culture continue to be depicted today (Bechdel, *The Indelible Alison Bechdel* 27–28). The key characters at the outset of the ongoing narrative included Mo, a stalwart lesbian-feminist; her best friend Lois, a vigorous sex-radical; and Clarice and her lover Toni, upwardly mobile law student and accountant respectively. This circle gradually expanded to include many other characters over the years, including, among others, new-age bi-dyke Sparrow; Ginger, English PhD student and eventually academic; women's bookstore owner Jezanna; stay-at-home-dad Stuart; Toni and Clarice's son Raffi; and queer theorist and compulsive consumer Sydney.

This chapter will particularly focus on the trade paperback collections that fall within the period under primary study: *Post Dykes to Watch Out For*, *Dykes and Sundry Other Carbon-Based Life-Forms to Watch Out For*, and *Invasion of the Dykes to Watch Out For*. Each collection reprints strips that have been previously published in newspapers and magazines (and since 1999 online) and adds an original longer sequence at the end of each volume. However, there will also be discussion of earlier strips, and the readings will be placed within their shifting cultural, thematic, and narrative contexts.

Dykes to Watch Out For creates a unique historical documentation of lesbian cultural and political history. This is due not only to the unusually long time over which it has been produced but also to Bechdel's conscious effort to be "somebody who traces the ebb and flow and flux of culture" (Bechdel qtd. in Cole). This is accomplished through her characters, who frequently discuss the pressing cultural or political issues of their day, and their multiple perspectives and arguments that prompt the readers' engagement with and discussion of many issues: trans inclusion in women's events; gay and lesbian mainstreaming; the book publishing industry; drag kinging; globalization; changing sexual mores; war; the position of lesbians in society; and government policy on a variety of issues. This occurs not only through the issues that are explicitly discussed through the characters and the narrative but also through the inclusion of background materials such as newspaper headlines, radio and television programs (sometimes taken verbatim from their sources, sometimes wittily tinkered with), or the shifting titles of books in the Madwimmin bookstore where Mo, Lois, and Jezanna work, which creates a pictorial record of lesbian texts, trends in gay and lesbian magazines, mainstream attitudes, and political debates.

Reading back over old issues of *Dykes to Watch Out For* can give a lesbian of my generation[1] a small, very accessible (and admittedly fictionalized) glance into a lesbian world filled with feminism, women's cafés and bookstores *and* the sex wars, leatherdykes, and lesbian mothers. It also gives a sense of the number of similarities that remain; articulates the multiplicity of lesbian cultural history that rejects the single-thread narratives that cultural histories are keen to promulgate; and ultimately gives a certain insight into the humanity of these lesbian characters, an attention to which would perhaps lead to less steadfast assertions of rebellion or of proclamation of being "Not Your Mother's Lesbians."[2]

Academic Reception of *Dykes to Watch Out For*

Critical attention to Bechdel's work has mostly been in the form of reviews and interviews, although Kathleen Martindale, Gabrielle Dean, Tuula Raikas, and Anne Thalheimer have undertaken academic work that examines *Dykes to Watch Out For*. The dearth of writing on the text has probably had less to do with its interest as a subject of academic attention than with its being a comic, a "low culture" genre associated with ephemerality. The academic study of comic books and strips has also been limited by the traditionally divided nature of academic disciplines, which renders comics, with their intersection of narrative, graphic art, and politico-cultural critique, difficult to place within a particular field. The fusing of disciplines undertaken by cultural studies and the current fashion for interdisciplinarity, together with the growing acceptability of analysis of cultural products that do not constitute "high culture" (of which the emerging field of pornography studies is a good example), means that perhaps further attention will be paid to these largely overlooked texts. There is also the matter of *Dykes to Watch Out For* being such a long-lasting and ongoing narrative. It can be somewhat intimidating to work with an ongoing text, as the conclusions that one reaches may cease to be relevant to the work as it continues to evolve. Further, the twenty years of ongoing production provides an overwhelming volume of material upon which to work.

In all the discussions of *Dykes to Watch Out For* that I have noted, except that of Raikas, the text is examined in relation to other lesbian comic strips—in each example Diane DiMassa's *Hothead Paisan* but also the work of Fish (Dean) and Roberta Gregory's Bitchy Bitch and Bitchy Butch characters (Thalheimer). While Dean examines lesbian comic strips' relationships with the phallic, Martindale, within the context of her monograph *Un/popular Culture: Lesbian Writing After the Sex Wars*, appraises *Hothead Paisan* and *Dykes to Watch Out For* in terms of their ability to be placed within the framework of a neo-avant-garde. Raikas addresses humor in *Dykes to Watch Out For* and Thalheimer engages with questions of multiplicity in sexual and gender identity framings.

Dean, reading dyke comic strips by Bechdel, DiMassa, and Fish through a Lacanian lens, asserts that "Bechdel suggests a dyke body politic that is fragmentary and composite, a social vision in which the 'phallus' is the organizing unity of dyke utopia—but is predicated, tellingly on disavowal." Dean argues that this occurs through the inscription of fetishes "that operate . . . as the phallus, illustrating the ambivalence of the distinction between 'having' and 'being' the phallus that subtends the normative sex and gender systems in the Lacanian scheme" (Dean 208). Dean is very critical of Bechdel, certainly more so than she is of her other subjects under consideration, asserting that

> *Dykes to Watch Out For* attempts to rewrite the Law of the Father as dyke domesticity in which difference, as a vector of unequal access to signification, ultimately disappears; the phallus here is nowhere visible as a bodily projection but is the organizing principle of community, which uses the portrayal of difference, paradoxically, to signify sameness. (211)

This "sameness" Dean claims is imposed precisely via Bechdel's studious attention to providing a diversified cast of characters:

> The careful inclusion of all "kinds" of dykes, along with Bechdel's illustrative attention to body size and race, as portrayed by hair and facial structure, results in a kind of compendium of dyke typologies. The well-intentioned diversity of the characters nevertheless produces a cast of stereotypes: none of them is complete in her own right, but together they produce a puzzle-picture of the community of choice. . . . the price of unity is the disavowal of difference, here racial difference—a standard foil for the deflection of anxiety in places where gendered hierarchies and normative sexuality are questioned. (212)

Dean is right that the literally two-dimensional characters she is discussing do include stereotypical elements. Each has identifiable characteristics—Mo is anxious, Clarice is terrified of commitment, Sydney is avaricious, Jezanna is bossy, and Lois is lusty. These clear character traits are necessary to add to the humorous dimension of the text, to organize the narrative, and indeed precisely to exposit differences—political, sexual, and philosophical—between the characters. To posit these stereotypical elements as rendering the characters as stereotypes, however, ignores the complexities and contradictions of Bechdel's characters.

Nor are these characteristics entirely unyielding. None of the characters are "complete in their own right," not only because we do not see a complete daily portrait of their outer and inner lives, but also because they *are characters* rather than complete human beings. Dean's perception that *Dykes to Watch Out For* is rewriting Lacan's mirror stage ("when a child sees in her reflection a unified, obedient and idealised body" 212) does not ring true, because the dyke culture portrayed in *Dykes to Watch Out For*, while certainly somewhat idealized, is certainly neither unified nor obedient. While many of the aggressive energies of the text are directed outward—at patriarchy, homophobia, or global capitalism—there are certainly conflicts between and within this community, and it is through these very conflicts that political or cultural issues are explored in a multifaceted way.

Dean does not fully articulate where precisely she sees a "disavowal of racial difference" in the text. Her only extrapolation of this difference is a brief comment that the "dyke family . . . is unified by political concerns, habits, values, a unity that would clearly be threatened by racial friction" (213) and the following:

> Racial equality within the scheme of the comic strip suppresses certain kinds of "difference" while pointing markedly to others: the characters have racially defined hair and facial features but not speech, and racial conflict is a constitutive problem of the outside world, emanating from it but not intruding on the dyke domestic. (213)

Bechdel indicates in *The Indelible Alison Bechdel* that her decision to give her characters the "racially defined hair and facial features" that Dean notes was a conscious one:

> I've never used any kind of shading to differentiate the skin color of my African-American characters. When I was starting to draw "Dykes," I noticed that a lot of white cartoonists, on the rare occasions when they included people of colour at all, used shading as the only way of indicating that a character was black. They would basically draw a white person, give them curly black hair, and fill in their faces with grey shading. So I tried to convey my characters' race by focusing on their features. Many of the shading styles I've seen other cartoonists use tend to obscure the characters' faces or seem prohibitively labor-intensive. (*The Indelible Alison Bechdel* 70)

To justify her conclusion that racial difference is obscured in the text, Dean gives only one example of the manner in which this occurs. Arguably, her criticism is founded upon a reading of the text that has both missed the moments of racial differentiation and conflict and presumes a universalized understanding that diversity itself is problematic and that a degree of political or social unity implies total inattention to structures of privilege. On the contrary, *Dykes to Watch Out For* is frequently and consciously engaged with just such structures of privilege.

In a recent online interview of Bechdel by *SistersTalk*, the interviewer states, "I enjoy and respect the way you use DTWOF to deal with racism and classism in the lesbian community. Those subjects are not always easy to talk about" (Stevens). Bechdel responds in a typically self-effacing manner, declaring that this reputation is "unfounded," as she has only "addressed racism in an indirect way by including a lot of non-white characters in my strip, and working hard to make them three dimensional people . . . But it's not the same thing as taking on racism as a topic" (Bechdel qtd. in Stevens). In comments posted in response to the interview, the interviewer observes that she thought Bechdel's comment was "interesting," as "[h]er work addresses both issues [racism and classism] in what I think is a very direct manner" (Stevens), while a reader responds with the same point, citing various examples where these issues were addressed. Stevens also observes within the interview itself that "Mo's character reminds me of someone who studies Black feminism. She seems to pay close attention to race, class status, gender, economics, sexuality, and culture—and how all these things reflect a woman's place in society" Stevens).

In her discussion of pictured racial diversity in Fish's work, Dean basically notes the same things as she does in Bechdel's work but comes to a radically different analysis. Analyzing Fish's scene in "One Fine San Francisco Evening," which shows two African American women and two Caucasian women having sex in different combinations, Dean reads that "the obvious absence of certain characters from the mix-and-match cast . . . reveals that the scene is not an exhaustive depiction of dyke sex, or even dyke S/M sex. Instead, racial difference provides only one among many sets of differences to be rearranged within the parameters of pleasure in this comic strip." (218)

Whereas Bechdel's characters are pronounced "stereotypes," Fish's characters are described as "merely outlines, to be filled in by variations" (217). What produces this difference in terminology and conclusion? The only explicit reason that Dean gives is the "obvious" absence of certain characters (examples given include Asian American, Latina, and disabled characters or femme bottoms)—that is, less diversity. The racial politics of DiMassa's work is also not seen as an issue, presumably because it is primarily composed of white characters, despite frequently "addressing" racism. Dean's argument thereby implies that it is *racial diversity* itself that is inherently problematic because it is seen to obscure *racial difference* and promote "sameness." That Bechdel's over two-decade-long narrative has undeniably provided "many sets of differences" among her characters—racial, sexual, ideological, and gendered—is seemingly not worthy of mention, and her depictions are understood to be intended as exhaustive. The selection of readers' letters to Bechdel included in *The Indelible Alison Bechdel* requesting new characters of various races, religions, disabilities, ages, interests, and sexualities seems to indicate that many readers don't see her cast as proffering itself as exhaustive. Martindale also calls attention to the racial politics of *Dykes to Watch Out For*. Writing of information gained through a telephone conversation with the author in 1990, Martindale remarks that

> Bechdel, a white, recovering Catholic lesbian, has apparently not been criticized for being a racist through under-representing lesbians of color or through arrogantly appropriating their cultures. She herself is not sure why this hasn't happened, because she acknowledges her own struggle against racism plays itself out in her difficulties with drawing lesbians who are different from herself. Bechdel thereby suggests how the struggle to know and love the apparently self-same is fundamentally entangled with the struggle to delineate social differences. When she began to draw her lesbians of color, she says that she depicted them more like the central white character's "ethnic sidekicks" than as fully fleshed lesbians in their own right. Nonetheless, even after five volumes, the center still belongs to the dominants within "the" lesbian community, rather than to leather dykes, butch daddies, or femme tops. While Bechdel's strip has interracial luppie moms, twelve-stepping Asian-American new agers, a cute young queer girl, and a disabled dyke, they're all second bananas. The lead is still Mo. (63)

I find it interesting that both Dean and Martindale switch so easily between discussions of the representation of racial diversity and sexual diversity (the latter being understood to mean the depiction of the "new" sex radicals) as if one can prove a point about the other—a feature that is not only problematic but also, I would argue, common to many sex-wars era texts having more to do with a certain generational disavowal of cultural products considered to be lesbian-feminist than a real concern with racism or a lack of sexual diversity in these texts. Ten years later, the lead in the comic strip is less clearly Mo, as other characters have begun to take more prominent roles in the series. Nonetheless, surely these two critics would not have considered it preferable for Bechdel to have made her lead character Mo (who is frequently cited as being based on Bechdel herself) a lesbian of color or to have drawn a less racially diverse cast of characters (as Dean appears to suggest); such approaches would undoubtedly have likewise drawn criticism for "arrogant appropriation" or racism respectively.

Martindale also views the characters as "stereotypes" while acknowledging that this may only be visible to "self-conscious lesbians and gays" (67). Despite her misgivings, however, Martindale is certainly more enamored of *Dykes to Watch Out For* than Dean. To Martindale, "[t]he strip . . . takes the form of an extended and unresolvable debate about sexual-political practices in a self-consciously lesbian-feminist subculture" (56) in which "Bechdel's chief subject is diversity between and among lesbians after the sex wars" (63). Again, while this may have been reflective of the early years of *Dykes to Watch Out For* (Martindale's work was written sometime before 1994), the strip has shifted along with the subcultures it reflects, and other subjects have overtaken this issue in prominence. In recent years, perhaps the most prominent issue in *Dykes to Watch Out For* has been lesbian mainstreaming and its contradictory yet contemporary companion, the increasing power and influence of the religious right.

Raikas, in "Humour in Alison Bechdel's Comics," takes quite a different approach, examining Bechdel's work through the lenses of the theoretical discourse surrounding humor. Raikas outlines the approaches to humor from theories of incongruity, dialogic theory, and feminist theory, identifying the manner in which Bechdel's comics are demonstrative of each form. An important aspect of humor theory Raikas outlines is that, at least "[t]o some extent, humour and joking are based on ingroup and outgroup relationships" (16), thus "[h]umour can be regarded as a product that results from and maintains group solidarity" (15). In the case of *Dykes to Watch Out For*, the "ingroup" that the narrative and thence humor creates and is also addressed to is formed by those more commonly found on the margins of both dominant narrative and humorous forms. The creation of such an "ingroup" thereby reverses hegemonic textual power relations. While this ingroup in Bechdel's work was initially solely comprised of lesbians, it has always addressed its comic attack at broader targets than homophobia and heterosexism, and the ingroup has expanded over the years to be inclusive of various other "carbon-based life-forms." In *Dykes to Watch Out For*, the aspect of the humorous exchange between author and reader that "maintains group solidarity" is particularly significant because it allows Bechdel to be much more critical of various aspects of lesbian cultural practices than would perhaps otherwise be deemed acceptable. Whereas in products such as *The L Word*, ostensibly "lesbian" jokes fall flat, *Dykes to Watch Out For* deploys multiple forms of humor that demand subcultural knowledge and do indeed succeed in producing real humor. This is due not only to the more able and witty writing skills of Bechdel but also to the manner in which the text positions itself in relation to lesbian communities and readers.

Raikas' analysis leads her to a somewhat different conclusion about *Dykes to Watch Out For* than Dean or Martindale; she sees Bechdel's cartoons as indeed representing difference via containing "a multitude of themes and issues. . . . From a feminist perspective, such a diversity shows that there is not just one kind of female identity and female experience, so to say one concept of woman, nor one way of existence that constitutes identity" (Raikas 68).

The rejection of both binaries and the fixed definition of terms and identity constructions (including lesbian) forms the thesis of Thalheimer's dissertation "Terrorists, Bitches, and Dykes: Gender, Violence and Heteroideology in Late 20th-Century Lesbian Comix." Thalheimer argues against various forms of coming out and "labeling," seeing these as creating binaristic hierarchies and instead pointing toward "framing." For Thalheimer,

> [t]o be framed as something . . . is more to be read as something rather than revealed as such because framing is not so concerned with the fixity of a single category. It allows the subject to be many things simultaneously, even contradictory things, without being permanently fixed to any of them, and grants the subject the ability to invoke and revoke frames at will. (10)

While this is an admirable thought, the extent to which "framing" can be applied in the world outside of theory and whether its provisionality and multiplicity can be maintained is questionable. It is also open to discussion whether such "framing" is entirely distinguishable from coming out, which may indeed recognize that a subject has multiple subjectivities and that sexuality is not the primary (or even *only* as Thalheimer at times seems to suggest) defining characteristic.

Indeed, Thalheimer herself does not make completely clear how one is to distinguish between "framing" and "labeling" and what the productive qualities of such a distinction are. The following is an informative example:

> By contrast [to outing oneself as a lesbian writing lesbian theory], being framed, or choosing to frame oneself, as a lesbian theorist (whether one identifies as such or not) increases lesbian visibility without the pressure of having to come out as anything. Invoking a lesbian frame can also grant the non-lesbian theorist a greater legitimacy. . . . In dealing with framing, [the nonlesbian scholar writing about lesbian subjects] is not forced to choose because no binary of lesbian/not-lesbian exists. She is already multiply framed as a scholar, as someone interested in lesbian theory, as someone who writes, as a woman, and so on. Nor do these frames assume any necessary hierarchical order or single predetermined identity. (13)

Such statements as "[i]n dealing with framing . . . no binary of lesbian/not-lesbian exists" are suggestive of a certain postmodern understanding of the world that sees the dissolution of "the binary" as an inherently progressive force. The proposition that "these frames [do not] assume any necessary hierarchical order or single predetermined identity" is a little strange, as readers may in fact construct their own hierarchy from this list according to interest or sequence. Indeed, Thalheimer herself has created a hierarchy by *removing* sexual identification from the list (unlike the other identity formations, it is seen as unimportant, ignoring that lived experience as a lesbian/queer woman may in fact be useful to writing lesbian theory) while fixing the single predetermined identities of "scholar" (some may be activists, journalists, writers, and so on) and "woman" (which may not necessarily be the case and is indeed itself a potentially binaristic term).

What bearing does all this have on *Dykes to Watch Out For*, a text explicitly framed by its lesbianism? Thalheimer proposes in her introduction that Bechdel's work is "often bitingly political, sharply satirical, and, most interestingly, highly infused with heteroideology" (26) and "highly heteronormative even while being very lesbian-focused" (30). Such statements would seem to imply that Thalheimer, like Dean, is very critical of Bechdel's work. Yet, within a chapter discussing Bechdel, Thalheimer appears to praise Bechdel for a story line that "critically reframes the ways we're taught to think about gender identity and sexual orientation" by "having Sparrow come out as a bisexual lesbian who's in a relationship that looks heterosexual under our learned cultural assumptions (seeming man + seeming woman = seemingly heterosexual relationship), even though all those terms are just as open for interpretation" and thereby allowing "talk about how identity and self-definition can shift and how certain terms—like gay and straight, and even bisexual—can be limiting, reductive, and, ultimately, non-descriptive" (86). Leaving aside for a moment the story line itself and what it achieves, can any "critical reframing" of normative ideas of gender and sexuality really be heteronormative or even a manifestation of heteroideology?

In *Come as You Are: Sexuality and Narrative*, Judith Roof defines heteroideology as "[t]he reciprocal relation between narrative and sexuality [that] produces stories where homosexualities can only occupy certain positions or play certain roles metonymically linked to negative values within a reproductive aegis" (xxvii). Roof also comes to the conclusion that "narrative peopled by lesbians still doesn't inscribe a different narrative or express a 'non'-heterosexual experience in a 'non'-heterosexual metaphor. . . . [W]e have difficulty making sense of same-gendered sex because sex is already heteronarrative" (122–23). In this work Roof cannot conceptualize lesbian narrative, or even orgasm, outside of a "heterosexual matrix," and even the basic physical properties of "cause-and-effect" (123) are given a heterosexual visage via Roof's theorizing.

Again, as in Dean's work, in Roof's analysis of a lesbian erotic story, any vestige of difference reads as negative, in this case heteronormative, although two given organisms, especially organisms as complicated as human beings, will necessarily be at least somewhat different to one another. Although heterosexuality is culturally inscribed as the dominant universalized sexuality, there are no psychological or narrative forces that compel a reader to read any given difference between a lesbian couple as heterosexual or influenced by heterosex should they be uninterested in reading it that way. Rather, the force that compels Roof's designation of this lesbian erotic story as an example of heteroideology is an ideological one: the psychoanalytic framework.

Roof's theory of the inevitability of narrative heteroideology is a classic example of queer theory's usage of the discourse of psychoanalysis. By deploying an ideology that is fundamentally heterosexist and misogynist despite multiple and various revisions and attempts by feminists and queer theorists to reclaim it, Roof becomes trapped in a paradigm wherein she can only view a text through psychoanalytic means and thereby can only produce an analysis that has less to do with the text itself than the theory she is using to examine it. These are the kind of insights on the basis of which feminist film theory for many years could not imagine an active female viewing position without recourse to "trans-sex identification" (see, for example, Mulvey, "Afterthoughts" 33). If one views actual performances, actual women, and actual narratives without a preconceived conception of where they fit within a very particular framework that is largely based on suppositions treated as unyielding truisms (and that it should be added, has been multiply asserted but is as impossible to *prove* as the existence of God), it can be seen that these inescapable necessities of analysis are perhaps not inescapable at all and are possibly, even completely, unnecessary.

Perhaps the best example of how such an ideology can influence the reading of a text is Thalheimer's deployment of Roof's theory in her analysis of *Dykes to Watch Out For*—a thoroughly and irredeemably lesbian text that, if anything, privileges lesbianism. Within Thalheimer's analysis, as was noted earlier, Bechdel's work becomes both "highly infused with heteroideology" (26) and "highly heteronormative even while being very lesbian-focused" (30). Reading *Dykes to Watch Out For* without the lenses of Roof's framework, one cannot begin to imagine how precisely a text *about* a multiplicity of lesbians, written by a lesbian, that has sustained a narrative for over twenty years despite including only the odd heterosexual coupling; where "normal" reproduction ironically involves artificial insemination;[3] and where there are multiple forms of lesbian relationships and sexualities on display can possibly be "highly infused with heteroideology" (26) and "highly heteronormative" (30).

THEORIZING LESBIANISM/LESBIANIZING THEORY

The reader may ask why I have engaged in this prolonged account of theoretical engagement with *Dykes to Watch Out For*. The criticism that exists upon *Dykes to Watch Out For* displays some fascinating examples of where theoretical excess and inattention to the text itself can lead to an interestingly telling misreading of a text. Added to this, such theorizations form an important part of *Dykes to Watch Out For*, which has over the years been very much engaged with the theories and ideas that have gained popularity within lesbian communities and academic circles. While in the early years this engagement was primarily with the ideological credos of lesbian-feminism, the introduction of the character Sydney in 1995 ("Poetic Justice" #228, *Hot, Throbbing Dykes to Watch Out For* 22–23)[4] signals a shift toward queer theory and *Dykes to Watch Out For*'s affectionate critique thereof. Sydney is in many ways both a parody of the pretensions of queer theory and an acknowledgment of its seductive potential. Initially a sparring partner for lesbian-feminist Mo, the two get together during the long final sequence of *Hot, Throbbing Dykes to Watch Out For* ("Sense and Sensuality" 95–141).

Sydney publishes in journals like *JLQT* (footnoted as the *Journal of Ludicrous Queer Theory; Post Dykes* 45), theorizes psychoanalytically about the famously homophobic Dr. Laura as "the firm parent, the wrathful deity, the dominatrix we all yearn for and fear" (#341, *Dykes and Sundry Other* 17), and writes articles with titles like "The Phallus Unzipped: Clinton's Dick and the Detumescence of Oppositional Subjectivity" ("Enough Already" #298, *Post Dykes* 8).[5] Sydney's lack of a world outside her research, outside theory, is parodied during her writing of "The Phallus Unzipped," when Mo reacts to the subject of Sydney's research paper by exclaiming, "Sydney, economies are collapsing all over the world! The Taliban has the women of Afghanistan under virtual house arrest! Refugees from Kosovo are freezing in the mountains! What about the *big picture*?" A perspective that Sydney demonstrates her complete disinterest in by responding, "Good point. I'll surf around and get some international perspective on the cigar incident" (#298, *Post Dykes* 9).

This overindulgence in the world of often-abstract theorizing is, however, presented not merely as worthy of parody but also as inherently seductive and filled with possibility. This is perhaps most clearly presented the morning after Sydney's first sexual encounter with Mo, when Sydney reappears in the bedroom carrying a cup of coffee and wearing a strap-on dildo. Sydney responds to Mo's statement, "Uh . . . I thought you said you were a femme" with "Oh, please, Mo. You're so provincial. Can't you see I'm disarticulating the epistemological foundation of gender through deferral and deconstruction of fixed sexual signifiers?" (*Hot, Throbbing Dykes to Watch Out For* 138). In the next panel, we see Mo, finger crooked, saying, "Well in that case, come here and do me, you big theoretical stud"—a statement that may be surprising to the regular reader in light of Mo's prevailing lesbian-feminist beliefs. The possibilities offered by the "deferral and deconstruction of fixed sexual signifiers" are not only apparent in making "available a different way of seeing that allows readers to recognize not only the meanings that they make, but also the meanings that they might *un* or *re*make by reinvesting new meanings into sexual practices" (Engel 86). Such deferral and deconstruction of fixed sexual signifiers also offers the wider possibility for imagining a world precisely outside of the "heteroideology" of which Roof speaks, a world where lesbianism, and lesbian sex, can be conceptualized without recourse to heterosexuality.

The intense yet fractious relationship between Sydney and Mo can also be metaphorically read as representative of the relationship between queer theory and lesbian-feminist thought. In *Identity Poetics: Race, Class and the Lesbian-Feminist Roots of Queer Theory*, Linda Garber persuasively argues against

> the by now axiomatic gap between lesbian feminists (sometimes more simply called "lesbian scholars" or "lesbian studies") and queer theory. I argue, first, that the divide, such as it is, is not, nor does it need to be, generational; second, that the divide, in certain ways, is *not* a divide but rather something like a failure to communicate, coupled with the coercive, divisive pressure of the academy to smash our forerunners—even if it means misreading and/or misrepresenting their work—to establish our own careers, which others will later take apart; third, that the defensiveness on the part of the "older" generation, which amounts to a rejection of new ideas, isn't real productive either; and, fourth, and centrally, that the creation and sustenance of the debate relies on the marginalization of working-class/lesbians of color. (5)

Sydney's citational practices are largely based upon Foucault, psychoanalysis, and other continental philosophies, and as such, together with the institutionalization of queer theory, she gains authority in academic circles (though despite this she must still scramble for tenure). Sydney's academic counterpoint within *Dykes to Watch Out For* is long-term character Ginger, whose thesis is entitled "Historical trends in dominant culture criticism of African-American literature, with a focus on the critical response to the influence of the oral tradition in contemporary fiction by black women." Both the difficulties of the contemporary academic climate and the differing statuses of various theoretical approaches within the academy are shown in the differentiation between the three lesbian scholars portrayed in the narrative, the third being Sydney's arch-rival for tenure Betsey Gilhooley, whose work appears to be infused with Marxist sensibilities. While Sydney gains tenure at their local university and Betsey is offered a position at Harvard, Ginger, after a dispiriting search for

Figure 6.1 "Notes on Camp" (#368, *Dykes and Sundry Other* 70–71)

Figure 6.1 (*continued*)

employment, is relegated to teaching uninterested students English at the less prestigious Buffalo Lake State College. Audre Lorde, acclaimed African American lesbian poet and perhaps Ginger's most treasured research specialization, is, interestingly, precisely one of the figures that Garber identifies as a central yet marginalized linkage between lesbian-feminism and queer theory. In the struggle for academic authority, here in the form of institutional positions, Ginger, and thereby Lorde's work, are marginalized and relegated to a place without status or influence.

In "Notes on Camp" (#368, *Dykes and Sundry Other* 70–71), Bechdel presents a lighthearted introduction to camp, and its redeployment of language (see Figure 6.1).

I have reproduced this strip in full because it is an excellent example of *Dykes to Watch Out For*'s engagement with the debates taking place within queer studies and of the manner in which Bechdel presents these debates in an accessible manner for a nonacademic readership. As one of the only strips in the series that is predominately peopled with male characters, it acknowledges the male-centric discourse of camp that has only recently begun to be transformed. It also points to where camp can fail and the importance of context. Carlos' "ironic and transformative nod" to the shirt known by Americans as a "wifebeater" out of its immediate context becomes problematic to two feminist parents attempting to bring up a feminist son. It is not only that Carlos' intentions are misinterpreted, but the potential for misogyny in various manifestations of camp can be seen in this depiction.

Lesbian Borders

The ongoing issue of acceptance of transsexual people into lesbian communities has been taken up in various ways in *Dykes to Watch Out For*. The first instance occurs through the presence of Jillian, a transsexual lesbian and a member of Lois's lesbian avenger group, who submits her work to "Madwimmin reads," a reading series of local lesbian writers Mo organises ("Au Courant" #193, *Unnatural Dykes* 52–53). Mo, as the icon of lesbian-feminism within the narrative, voices her trepidations and prejudices at the concept of a transsexual lesbian reading in the event before discussing the issue with Lois and eventually finding herself defending Jillian to another woman at the reading who questions Jillian's right to signify herself as "lesbian."

Later on in the series, overhearing Mo ranting about her "male energy" and proclaiming that "[a]ny day now, our friend Lois is gonna saunter in here and tell us to start calling her "Louis"!" (#351, *Dykes and Sundry*

Other 37), Lois decides to cure Mo of her negative attitudes toward FTM transgendered people by pretending to be transitioning to male. This segment continues for much longer than the first engagement and includes much discussion of the issue as well as a scenes in which Mo herself is mistaken for a man, which suggests the potential for camaraderie between those whose appearance does not necessarily signify their gender (#359, *Dykes and Sundry Other* 53). In "Girls! Girls! Girls!" Mo comes to terms with Lois's decision to transition to male, after which Lois informs her, "I'm not transitioning at all . . . I enjoy being a girl. In a perverse kind of way" (#367, *Dykes and Sundry Other* 69). This strip is then used to introduce an ongoing real trans character into the narrative.

With recent court decisions allowing hormone treatments for teenagers and even an *Oprah* episode featuring transgender children ("The 11-Year-Old Who Wants a Sex Change" first screened 5 Dec. 2004), the transgender identities of younger people are in the process of becoming prominent issues. This has been explored in *Dykes to Watch Out For* via the character of Jonas/Janis. First introduced as a ten-year-old in the 2001 pride parade strip ". . . Before a Fall" (#365, *Dykes and Sundry Other* 64), Jonas is shown in the next strip to be somewhat female (notably after the strip in which Lois confesses to Mo that she is not really transitioning), which is identified by her insistence during a game with the adult characters that "we're not guys, we're Powerpuffs . . . I'm Blossom" (68). While Lois encourages Jonas with gifts of female clothing and attempts to provide her with "a genderqueer role model" (124) and Sparrow introduces Jonas as "Jonas—who prefers to be addressed as Hermione, and who's probably transgender" (141), Stuart doesn't "think it's good to encourage Jonas in his . . . his gender confusion. You're not doing him any favors. It's a harsh world out there, and eventually he's gonna have to *face* it" ("The Awful Truth" #406, *Invasion of the Dykes* 23). Stuart's insistence on facing "the awful truth" is expeditiously narratively undermined by his quickness in trying to protect Jonas from the similarly awful truth of institutionalized racism.

Two years later, Jonas is officially known as Janis and attends a summer camp for transgender children ("Siren Thong" #446, *Invasion of the Dykes* 102–3); a further year later, Janis begins to engage in trans activism by attending Camp Trans at the Michigan Women's Music Festival with her mother Jasmine and Lois ("In the Good Old Summertime" #471). Bechdel notes in a recent interview that "I'm [keeping current with issues in the gay community] with this young trans character Jani[s], who's 14," indicating that "[w]e're going to have this big discussion about whether she's going to start hormones" (qtd. in "Hothouse Talks to").

It is significant that, in both Mo and Stuart's case, concerns and prejudices are allowed to be voiced before coming to a fuller understanding and acceptance of various manifestations of transexuality. This is a very different manner of exploring the issue than occurs in, for example, the 1999 lesbian film *Better than Chocolate*, which portrays a very sympathetic transsexual lesbian character who is assaulted by a lesbian who is simultaneously signified as lesbian-feminist, emotionally unstable, and violent. *Better than Chocolate*'s strategy of convincing its audience through a hyperbolic denunciation of a lesbian ideology and generation—much like *Queer as Folk* and *The L Word*'s strategies of convincing their audiences of perspectives regarding assimilation and art through outing—are rather simplistic. Though they have the possibility of being effective by forcing their viewers to identify with a particular character through presenting a larger-than-life violent or hypocritical character as their binary opposite, they do little to actually explore the issues and arguments involved or lead their viewers on a journey of education, relying instead on simple shock factors. In *Dykes to Watch Out For*, readers are given the opportunity to identify with Mo and, in the latter case, with Stuart, in their concerns and arguments, which can then be countered by explanations, arguments, and the recognition that "who am I to question someone else's identity?" (Mo in "Lime Light," *Unnatural Dykes* 54).

It is not only the borders between lesbians and other "carbon-based life-forms" that have been opened up in recent *Dykes to Watch Out For*; the very definitions of, and associations with, lesbianism have shifted to represent elements of new lesbian generational trends that are quite distinct from previous trends within, and understandings of, lesbian communities and subjects. Whereas Mo and her friends are quite clear about their identity as lesbians and comfortable and proud to announce the term, some members of the younger generation, represented by Madwimmin intern Sophie in *Post Dykes*, indeed, like Thalheimer, denounce the idea of "labels": "Why are there all these, like, separate categories? . . . Like, if I have a boyfriend or a girlfriend or *whatever*, why do people have to *label* me? It's so *shwag*, know what I'm saying?" ("Youth Wants to Know" #299, *Post Dykes* 10–11). Later, in *Invasion of the Dykes to Watch Out For*, we see the embodiment of a homocon in new character Cynthia. Where once lesbian consciousness was generally understood to represent certain leftist political beliefs in keeping with the oppressed status of homosexuals, Cynthia represents the paradox that is the conservative homosexual. Interning in a "right to life" office and learning Arabic to join the CIA ("Fundamental Differences" #441, *Invasion of the Dykes* 93), she takes some time to emerge from the closet and is deeply surprised when those who share her ultraconservative beliefs are outraged at her lesbianism. She continues to pursue her Republican projects, demonstrating her naïve assimilationist faith that if she behaves like they do, she will be accepted. Later, while campaigning for George Bush's reelection, she observes of the presidential debate, "When [John Kerry] called Mary Cheney a lesbian, I felt like I'd been *drop-kicked*," revealing the extent to which conservative assimilationists respond to the very word lesbian as if it were an insult, which demonstrates her closetedness despite being out at this point ("Absolute Value" #452, *Invasion of the Dykes* 114). A further disjunction between her beliefs and her reality is exposed by her "wait till marriage" virginity pledge ("The Temptations" #467) and concurrent support of the very Republican party that introduced such measures as the Defense of Marriage Act—something that attempts to disable her from ever getting married.

The significance of these representations in *Dykes to Watch Out For* is that each of these identities coexists, not only with other "new" identities that are now in the glare of media and theoretical spotlights, but alongside identities perceived to belong to an older generation, which is likewise diverse. Such diversity serves as a counter to narratives of fashionability that posit each lesbian generation or decade with a specific attribute and act as if each lesbian generation replaces rather than coexists with its predecessors.

MAINSTREAMING

Another key theme of Bechdel's work is mainstreaming, a topic of central significance to this monograph and a key issue in contemporary lesbian cultures. Here, the discussion does not so much circulate around accusations of assimilationism as can be seen in *Queer as Folk* (though some characters do accuse others of being assimilationist at times), but rather it centers on the larger question of what will happen to lesbian culture, previously existing as a quite separate and frequently oppositional entity to mainstream culture, as it begins to enter the mainstream. As Kelly Mack puts it,

> Bechdel's most recent interest and concern is about queer culture assimilating into mainstream culture with such phenomena as Ellen coming out on national television. She described "horror" and fear that the separate sub-culture will disappear and is yet "fascinated" that visibility and equality could erase the very people that fight for them. To this effect Bechdel related that her characters play out the clash between her desire to be an

"outsider" versus a "citizen."... Bechdel sees these characters work in collaboration as people in the queer movement work the assimilation problem from all sides.

Within the narrative of *Dykes to Watch Out For*, characters such as Toni fight for marriage equality, and queer theorist Sydney claws her way into the cultural recognition of a tenured place in the academy, while Mo, Lois, and Sparrow vigorously defend their outsider, oppositional status.

Over the years, those classic icons of mainstreaming—having children and same-sex marriage—have been both debated and practiced in *Dykes to Watch Out For*. Long-term couple Clarice and Toni celebrate their commitment ceremony in a 1990 strip ("Altared States" #87, *Dykes . . . : The Sequel* 26–27), have a Vermont civil union in 2000 ("Cry Cry Cry" #349, *Dykes and Sundry Other* 32), and get married at city hall in 2004 during a temporary legalization of same-sex marriage ("Get Me to the Clerk on Time" #436, *Invasion of the Dykes* 83). The ongoing nature of these ceremonies is suggestive of a different kind of narrative logic than that found in traditional heterosexual narratives in that the act of marriage is not a narrative endpoint but rather a belated

Figure 6.2 From "Get Me to the Clerk on Time" (#456, *Invasion of the Dykes* 83)

and ongoing process, while the celebration thereof does not legitimate sexual and social coupling but rather acts as a recognition of previously existing coupled states. Ultimately, this renders the very act of marriage more clearly a legal and social construct rather than a necessity for long-term bonding between individuals.

While Mo and Clarice may consistently argue over whether lesbian marriage is an assimilationist or a defiant gesture, Sydney puts it best when she proposes to Mo outside *Dykes to Watch Out For*'s fictional city hall (see Figure 6.2).

This single frame encapsulates the conflicting politico-ideological meanings of same-sex marriage, with Sydney's words reflective of Kath Weston's assertion that "ostensibly similar formal features of kinship can carry conflicting meanings and embed subtle ideological shifts, allowing 'new' family forms to be read simultaneously as radically innovative and thoroughly assimilationist. In the end, they are intrinsically neither" (64). The birth of Raffi, child of Toni and Clarice, graphically depicted in "Flesh and Blood" (*Spawn of Dykes* 99–132), marked the first child born within the *Dykes to Watch Out For* narrative. Jiao Raizel (JR), the child of bi-dyke Sparrow and sensitive-new-age-guy Stuart, initially appears to be born into a much more normative family structure due to the genders of her parents. Yet Sparrow and Stuart's communal living style and JR's unconventional upbringing, in comparison to Raffi's suburban home and nuclear family, question the automatic normativity of this family, once again outlining how the complexities in queer family structures can make the same family seem both "radically innovative" and "thoroughly assimilationist" (Weston 64).

In *Dykes and Sundry Other Carbon-Based Life-Forms to Watch Out For*, the saddest element of mainstreaming (and globalization) takes place in the form of the closure of Madwimmin, the women's bookstore that has formed the central communal space necessary for the development and maintenance of this subculture, its ideas, and its networks. This can be seen as analogous to the closing of Firebrand, the feminist publishing house that was responsible for the first nine volumes of *Dykes to Watch Out For* (though Firebrand has since been reopened), as well as the more general phenomenon of the closure of specifically "women's" and lesbian businesses and organizations. Madwimmin closes due to not only the increased pressure of major bookselling corporations such as Amazon.com and Barnes and Noble (known humorously within the narrative as "Medusa.com" and "Bunns and Noodle") but also the increased presence of books once only available from subculture venues in mainstream bookstores that has been the result of lesbian and gay mainstreaming. While this is in many ways a positive development, it has added ramifications for the survival of women's and gay and lesbian bookstores and the kinds of cultural possibilities they allow.

The question of what kinds of cultural products mainstreamed culture—consistently aware not only of heterosexual opinion but also of the financial bottom line—will allow to flourish remains to be seen, though Madwimmin's proprietor Jezanna predicted doom in 1999: "Just wait till thousands of diverse, lovingly tended independent bookstores get replaced by generic chains, subliterate *bean counter* [sic] down at corporate HQ selecting all the titles! You can say goodbye to books by risky or unproven authors, like the ones I've been hand-selling our customers for years. We'll be up to our *cerebral cortexes* in 'Celine Dion's Titanic Cheesecake Recipes'!" (#307, *Post Dykes* 27). Upon closing the bookstore, Jezanna remarks, "When I opened this place 25 years ago, this store was an outpost in a hostile environment. The future was uncharted. I had no idea what I was getting into, except that it wasn't going to make much money." Lois replies, "What an achievement, to come full circle." Mo observes, "Jeez, I thought we were gonna make the world safe for feminism," to which Jezzanna replies, "We did. To be packaged and sold by global media conglomerates" ("Same as It Ever Was" #397, *Dykes and Sundry Other* 129).

While some venues display a simple, celebratory attitude toward mainstreaming, *Dykes to Watch Out For*, through its multiple characters, presents a variety of opinions about it, which allows the text to build a more complex picture of mainstreaming, complete with potential pitfalls as well as benefits. An early interaction that displays these multiple positions can be seen in "Double Play" (#249, *Hot, Throbbing Dykes to Watch Out For* 64–65). Viewing Ellen DeGeneres's jokes about being "Lebanese" before her official public outing, Mo complains of the coyness thereof, while Sparrow asserts, "It's network TV! They did more for lesbian visibility with one stale pun than you'll ever do" (65). It is left to the reader to determine whether Mo or Sparrow is right or to synthesize elements of both their arguments and thereby recognize the complex and multiple nature of progress for gay and lesbian rights.

The extent to which queer viewers may feel thoroughly disgruntled with the wholesale commodification of their lives by contemporary film is further explored in "Deductive Reasoning" (#389, *Dykes and Sundry Other* 112). Considering a film to see, Lois and her friend Jerry (a transsexual man) find *Sorority Boys* (2002) and *Kissing Jessica Stein* (2001) playing.

Both films contain elements readable as queer, the former through its cross-dressing motif, which is reminiscent of classic subtextually queer Hollywood films like *Some Like It Hot* (1959), as well as a story line wherein a young woman falls for one of the cross-dressed male protagonists who is then revealed to be male and thus recuperates the heteronormativity of the narrative. The latter centers on a straight woman, who, frustrated by her lack of male romantic prospects, dates a bisexual woman before an ending that reaffirms her inherent heterosexuality. Upon hearing these choices, Lois proclaims, "God, how totally depressing." The next frame sees her mood suddenly uplifted, and she asserts, "Hey, look! 'The Children's Hour' is on TV at eight. Let's just stay here." Lois's preference for older, pathologized depictions of homosexuality over the now-available trendy yet marginalized images acts as a critique of those ideologies that see any representation in popular culture as good representation and the history of homosexual liberation as a linear one toward progress.[6]

Various strips examine the unequal nature of mainstreaming, questioning whether it is indeed in effect for all gays and lesbians or simply those who have gained sufficient monetary or celebrity power to claim (buy) their "place at the table." In "Gimme Shelter" (#355, *Dykes and Sundry Other*), bi-dyke Sparrow is dining out with Stuart "slightly beyond their means." She comments, "I don't think I'll ever stop being amazed at the kind of service I get when I'm out with you. When I'd come here with June [Sparrow's ex-partner] we were lucky if they gave us water" (44). June, once a recovering alcoholic, now an internet entrepreneur, is separately at the restaurant and comes over to tell Sparrow, "Remember how rude they used to be to us in this place? Now I eat here every day 'cause my office is

Figure 6.3 From "Intro to Queer Theory" (#399, *Invasion of the Dykes* 9)

upstairs, and they *love* my ass" (45). Captivated by June, the waitress then ignores the lower-income Sparrow and Stuart. In this sequence we see homophobia transformed into acceptance—an acceptance that is clearly based around financial considerations.[7] This is then reinforced in "Intro to Queer Theory" (#399, *Invasion of the Dykes*). Bechdel's strip starts with the extradiegetic comment "The arc of history is long" before going on to depict a narrative that sees Raffi coming home from school having been engaged in a game of "Smear the Queer" (8). Another aside announces, "but it bends towards justice," before Raffi"s teacher assures Toni that she will have a talk to the students about the game. The strip then goes on to conclude, "Sort of (apologies to Dr. King)" (see Figure 6.3).

This conclusion questions the kinds of justice history bends toward. While newspapers may now be willing to print same-sex commitment ceremony announcements and discuss gay issues, in the final frame, this is revealed to be at a price—queers fully participating in exploitative capitalist projects are represented here by the investment bank's "global rapine & pillage division." And once again, as in the situation of June, it is socioeconomic status that has allowed this mainstreaming.

These ambiguities of progress, however, do not mean that lesbianism is thoroughly mainstreamed. The ongoing tenacity of oppositional lesbian cultural productions such as *Dykes to Watch Out For* makes it highly unlikely that even the most concerted of mainstreaming politics or a veritable influx of glossy televisual lesbians will be able to stop either the production and distribution of such texts or the broadly politicized nature thereof.

CHAPTER 7

Conclusion

THE EXISTENCE OF *DYKES TO WATCH OUT FOR*, with its multiplicity of characters and politics, its historicizing function, and its simultaneous deployment of insider and outsider discourse, is situated here at the endpoint of my argument about contemporary lesbian cultural productions. This is due to my conviction that this variegated and historicized narrative not only presents the plurality of lesbian representations and communities—beyond that which fashion-conscious academic or media catchphrases like "queer culture" or "lesbian chic" can capture—but also allows and indeed encourages a rethinking of our modes of theorizing texts through the stringent lenses of specific theoretical modalities.

In the wake of the success of her graphic memoir *Fun Home* and due to her work on her new memoir, Alison Bechdel cut back on the number of *Dykes to Watch Out For* strips released each month in May 2007 and began reprinting archival strips on her online blog in between new strips. It quickly became apparent that these twenty-year-old strips were not nearly as out of date as one might think. As Bechdel noted when reprinting strip number eight on July 11, 2007, "It was twenty years ago today . . . and the times, they aren't a-changin'." This strip that observed the political situation of the time is almost completely applicable to the current political situation with the simple substitution of a politician's name here and there. And while television shows such as *Sex and the City* were keen to point out the movement of lesbians into power and the mainstream in 1999 (see further discussion in Hidalgo), Bechdel was observing this phenomenon in "Look Out! It's Luppies" back in 1986 (*Dykes to Watch Out For* 46), pointing toward the cyclical nature of culture.

Queer as Folk's lesbian characters are not adequately expressive of lesbian culture precisely because they exist as singular projections of a host of gay male attitudes toward lesbians that conflate and redeploy distinct elements of lesbian and queer cultures in the name of creating a binarized opposition between lesbians and gay men. In the final episode of the United States' *Queer as Folk*, the series acts to reassert its self-delineation as antiassimilationist through a speech given by Michael on behalf on the "Committee for Human Rights." Michael deviates from the script provided for him that emphasises that gays are "just like straights," instead emphasizing diversity, especially within "the gay community," and asking "do we all have to have the same lives to have the same rights?" While the message is admirable, it bears little resemblance to the visual and dialogic signifiers that *Queer as Folk* has in truth presented, which have been marked by their veritable lack of diversity. The series then ends with the bombed-out Babylon imaginatively transformed once more into a queer sociocultural space, complete with pounding disco beat. By reinscribing gay male club Babylon as the representation and epitome of queer culture in *Queer as Folk*, the series effectively displaces lesbians to a realm outside of queer signification.

Each of the texts I have discussed throughout *Lesbians in Television and Text after the Millennium* includes a strong visual element. The significance of the visual is nowhere clearer in contemporary culture than in *The L Word*, which subtextually enacts a curious redeployment of previous modalities of understanding lesbian culture without being figurally representative of them. This is most peculiarly presented in the series' attitudes toward femme lesbians, though negative modes of understanding butch-femme relationships, female masculinity, and nonnormative genders are also presented by *The L Word*. The series' definition of pornography as violence likewise hearkens back to pre-sex-wars attitudes within lesbian culture as well as the institutional deployment of the arguments that were seen in such circumstances as the *Bad Attitude* trial (see Cossman et al.).

If *The L Word* is willing to subtextually reinvigorate certain negative attitudes about particular lesbian gender presentations and pornography, then *Gurlesque* is a noughties representative of the cultural products that emanated from the debates that sought to repudiate such attitudes, and it blurs the boundaries between art and pornography that *The L Word* is so keen to promulgate. It also plays a reconciliatory role between positions characterized during political and academic debates as intrinsically opposed: lesbian-feminism and queer culture.

Perhaps such returns and syntheses are inevitable. A story Elizabeth Freeman recounts in "Packing History, Count(er)ing Generations" illustrates this point well. Freeman writes of "teaching my first solo course in Lesbian and Gay Studies in 1993, at a moment when "identity" was rapidly morphing into cross-gender identification" and finding her assumptions challenged by a younger graduate student who "dressed like my feminist teachers had in college" (727). This student's

> self-presentation ruptured any easy assumptions about lesbian generations and registered the failure of the "generational" model to capture political differences between two women who had race, class, nationality, and sexual preference in common. The temporal incongruity of her body suggested that she simply did not identify with what I would have taken to be her own emergent peer culture of neopunk polymorphs, Queer Nationals, Riot Grrrls, and so on—nor with my culture of neo-butch/femme, consumerist sex radicalism. Her body's "crossing," then, was different than the gender crossings that queer studies was just beginning to privilege. It was a crossing of time, less in the mode of postmodern pastiche than in the mode of stubborn identification with a set of social coordinates that exceeded her own historical moment. (727–28)

Freeman's anecdote is illustrative of the reality that there will always be those who exist in such temporal incongruity with generational trends, and when we posit a clear journey between one generation and the next, we write over such individuals. While one can locate, in a generalized manner, the particular features of a "lesbian generation," there will always be diversity within that very generation: those who identify with a time that has been, those who identify with a time still to come, and those who present something different altogether.

Together these disparate and seemingly unconnected texts reveal their connections—through contemporary cultural thematics such as the assimilation debate; through their continued deployment of the debates of the sex wars and relationship with feminism; through the often complex negotiations of signification between the immediately visible and subtextual levels of each text; and through their positioning as always negotiating the insider/outsider dichotomy that plagues contemporary lesbian cultural representations.

I have had a concern throughout the production of this monograph regarding its potential usefulness. To what purpose am I writing this? What do I hope to achieve? As Bonnie Zimmerman asks in her introduction to *New Lesbian Criticism*, "how do we make our theories useful?" (12). I hope that with this work I can at least gesture toward a certain multiplicity of theorisation and emphasize the importance of placing the analysis of mainstream cultural products in the context of existing lesbian cultural production and placing the texts themselves rather than an abstract theory as the central subjects of analysis. I hope to step away from theorizations of lesbian culture in terms of the new or the older to locate the multiple influences and significations inherent in each text; to encapsulate some essence of enormously variegated cultural production and the plurality of the debates that exist in lesbian communities; and perhaps to achieve some sort of reconciliation of those parties frequently presented as intrinsically opposed that are not always so different after all.

Notes

Chapter 1

1. Interestingly, Kera Bolonik, the author of the "Not Your Mother's Lesbians" feature article, is now the author of the "official companion book" to the series, *The L Word: Welcome to Our Planet*.
2. A significant dialogue about this generational oppositionalism exists (see, for example, Martin, "Sexualities without Genders" and Walters, "From Here to Queer"), and the phenomenon continues in the contemporary context. Recent examples include Bolonik's "Not Your Mother's Lesbians," Levy's "Where the Bois Are," and K. Fox's "Blush: Making Our Own Erotica" (all 2004).
3. The influences of academia on such cultural products as drag kinging are perhaps most thoroughly mobilized through the participation of academics and postgraduate students in drag king troupes and performances. Vicki Crowley writes of several academics and PhD students taking part in a drag king troupe in "Ben Dover and His Beautiful Boys" (288–89), and writers in the field such as Donna Jean Troka also perform in drag. An example of academics participating in and influencing lesbian performance culture can also be seen in the Chicago burlesque scene, with burlesque historian Tara Vaughan Tremmel recruiting performers for and organizing burlesque shows.
4. The full quote reads, "I hate that term 'in the closet.' . . . Until recently I hated the word lesbian too. . . . I've said it often enough now that it doesn't bother me. But lesbian sounded like somebody with some kind of disease. I didn't like that, so I used the word gay more often" (ellipses on interviewer asides).
5. Although I am characterizing the 2003–2004 focus on lesbians prompted by the release of *The L Word* in the mainstream media (occurring in the midst of a more general mainstream fashion for gayness as seen in the 2003 *Vanity Fair* cover story "TV's Gay Heat Wave") as the second wave of lesbian chic, one could argue that it is simply an extension or augmentation of early nineties lesbian chic.

6. *Gurlesque* often includes drag acts, *The L Word* briefly features the character Ivan Ay Cock, *Dykes to Watch Out For*'s Lois is frequently seen as her drag persona Max Axle, and *Queer as Folk* ever so briefly shows the characters watching a performing drag trio.
7. Examples include *Desert Hearts* (1985), *All Over Me* (1997), *Better than Chocolate* (1999), *But I'm a Cheerleader* (1999), and *D.E.B.S.* (2004). Many of these films do include a readable-as-butch lead role, but each of these characters displays some markers of femininity. *But I'm a Cheerleader* does include a very butch secondary character who is, however, later revealed to be heterosexual.
8. See, for example, the following quotation from Levy's "Where the Bois Are": "What all bois have in common is a lack of interest in embodying any kind of girliness, but they are too irreverent to adopt the heavy-duty, highly circumscribed butch role. To them, butch is an identity of the past, a relic from a world of Budweiser and motorcycles gone by" (n.p.). In *The Drag King Book*, Halberstam and Volcano also note a reluctance to identify as butch among the drag kings of New York (120–25), though they also note that "the Drag Kings in London almost exclusively identified as butch or transgender" (125).
9. Examples include Butler's *Gender Trouble*; Cahn's "From the 'Muscle Moll' to the 'Butch' Ballplayer: Mannishness, Lesbianism, and Homophobia in U.S. Women's Sport"; Inness and Lloyd's "G.I. Joes in Barbie Land: Recontextualising the Meaning of Butch in Twentieth-Century Lesbian Culture"; various articles in such anthologies as Munt's *Butch/Femme: Inside Lesbian Gender*, Halberstam's *Female Masculinity*; and Doan's "Passing Fashions: Reading Female Masculinities in the 1920s."
10. For further discussion of the category of the "feminine invert," see Clare Hemmings's "Out of Sight, Out of Mind?"
11. Essays that address lesbian chic include Clark's "Commodity Lesbianism"; Stein's "All Dressed Up, But No Place to Go? Style Wars and the New Lesbianism"; O'Sullivan's "Girls Who Kiss Girls and Who Cares?"; Pottie's "Hierarchies of Otherness: The Politics of Lesbian Styles in the 1990s, or What to Wear?"; Cottingham's short monograph *Lesbians Are So Chic . . . That We Are Not Really Lesbians at All*; Inness's chapter "'They're Here, They're Flouncy, Don't Worry about Them': Depicting Lesbians in Popular Women's Magazines 1965–1995" within her monograph *The Lesbian Menace* (52–76); and Ciasullo's "Making Her (In)Visible: Cultural Representations of Lesbianism and the Lesbian Body in the 1990s."
12. For further discussion of lesbian and bisexual representation on *Law and Order*, see Lo, "Lesbian and Bisexual Women on *Law and Order*."
13. During research in the Lesbian Herstory Archives in Brooklyn, I discovered the existence of groups such as the media committee of Lesbian Feminist Liberation, Inc. (New York, 1975; see Cotter,), the Gay Media Project (Philadelphia, date unknown, appears to be contemporaneous with other groups, c. 1970s), and the Gay Media Coalition (New York, 1975), which aimed to fight media bias against gays and coproduced and advised public affairs broadcasts on homophobia through encountering letters and press releases from these groups. The pioneering

efforts of these groups and activists were no doubt essential to introducing positive homosexual representation to television screens, and their role in creating the groundwork for the contemporary growth in lesbian and gay images on television must be acknowledged. Such organizations also prefigure the role now played by GLBT media watchdogs such as the Gay and Lesbian Alliance Against Defamation (GLAAD).
14. "Unpacking the L-Word: Lesbian Representation in Contemporary Culture" at the 46th Annual Midwest Modern Language Association (M/MLA) convention in November 2004. Although the call for papers asked for discussion of any contemporary texts including lesbians, seven out of the eight panelists discussed *The L Word* (see Beadling, Beirne "Lipstick and Lesbians"; Douglas "That's Not Me"; Jaarsma and Pederson; Renshaw and Robinson; Williams and Jonet; and Wolfe and Roripaugh.
15. Natalya Hughes has also observed this dimension of *Gurlesque*.
16. See, for example, Bensinger; Conway "Inhabiting the Phallus"; Katrina Fox "Blush"; and, to a lesser extent, Engel.

CHAPTER 2

1. There are subtle differences in popular understandings of what mainstreaming and assimilation constitute; undertaking a study and definition of these differences is, however, beyond the scope of this project.
2. According to Vaid, "[T]he antigay ban [in the military] was not an issue on the minds of most gay and lesbian Americans during the 1992 campaign: AIDS and the antigay [R]ight were" (161).
3. This excludes adoption by the couple and church weddings and stipulates that one partner must be a citizen of and reside in Denmark. A step back from this early progress was taken when, in 1997, assisted insemination for lesbians was banned (Jensen 39).
4. The law provides the "[s]ame rights and responsibilities as opposite-sex married partners" with the exception that the couple "cannot adopt a child from abroad" (ILGA-Europe, "Legal Recognition").
5. In Belgium's case, the exception is that "Belgian law does not provide for presumed paternity for the female spouse of a married woman who gives birth during their marriage; no provision for joint parental responsibility, nor for adoption by a same-sex partner or a same-sex couple" (ILGA-Europe, "Legal Recognition").
6. "Same-sex married partners will now enjoy all the rights and responsibilities of marriage, including entitlement for joint adoption" (ILGA-Europe, "Legal Recognition").
7. "Beyond Gay Marriage," a chapter in Warner's monograph *The Trouble with Normal*, provides a fuller account of the various positions and arguments in the debate over same-sex marriage (81–147).
8. Interestingly, it took Camper six years to find a publisher for *Juicy Mother*, her anthology of queer comics. It is now published by Soft Skull Press, and a second volume of *Juicy Mother 2: How They Met* was published in late 2007.

9. *Cheer Up!* (2004), a short documentary on radical cheerleading, was featured as one of the April 2006 Dyke TV segments rebroadcast online.
10. These struggles are discussed in English in the documentary *Dangerous Living: Coming Out in the Developing World* (2003).
11. While representations of homosexuals within the "married with children" model can be indicative of certain normalizing attitudes and political strategies, lesbian parenting in practice should not be automatically considered as a gesture of assimilation and indeed can be read as a radical remodeling of the nuclear family. See further discussion of this issue in Chapter 3.
12. Examples include Brill's *The Queer Parent's Primer: A Lesbian and Gay Families' Guide to Navigating through a Straight World*, Clunis and Green's *The Lesbian Parenting Book: A Guide to Creating Families and Raising Children*, and Snow's *How It Feels to Have a Gay or Lesbian Parent: A Book by Kids for Kids of All Ages*.
13. I acknowledge that the masculinity of many drag kings is not solely performative; however, it can be, so for my purposes here I am using this as an example.
14. An example of a cultural product growing out of this debate is Stacey's comic/'zine "Welcome to Sunny Camp Trans," produced by Butch Dyke Boy Press and the Boston Lesbian Avengers in 2000.
15. A central narrative force in the film is a fictionalized portrayal of the battle between the Little Sisters bookshop and Canadian customs that is frequently cited by propornography lesbians as the outcome of antipornographic feminist activism.
16. Further reading on the subjects of transexuality, transgenderism, and genderqueer can be found in *Gender Blending: Confronting the Limits of Duality* and *FTM: Female-to-Male Transsexuals in Society* by Holly Devor; *Invisible Lives: The Erasure of Transexual and Trangendered People* by Viviane K. Namaste; *Gender Outlaw: On Men, Women, and the Rest of Us* by Kate Bornstein; *Transmen and FTMs: Identities, Bodies, Genders, and Sexualities* by Jason Cromwell; *Lesbians Talk Transgender* by Zachary I. Nataf; *Trans Liberation: Beyond Pink or Blue* by Leslie Feinberg; and *Genderqueer: Voices from Beyond the Sexual Binary*, edited by Joan Nestle, Riki Wilchins, and Clare Howell.
17. Note that these texts (with the exception of Howes's) all engage almost exclusively with American television. While a significant amount of articles have been written on specific programs from other nations and various internet-based articles chart a genealogy of development, to my knowledge no large-scale study has been undertaken to that effect in English that is not significantly biased toward American examples to the exclusion of (almost) all else. A very good early (non-academic) article on lesbians on television in the U.K. can be found in Nicki Hastie's "It All Comes Out in the Wash: Lesbians in Soaps." A magazine article by Keith Howes, "Gays of Our Lives: 30 Years of Gay Australian TV," and my forthcoming "Screening the Dykes of Oz: Lesbian Representation on Australian Television" (*Journal of Lesbian Studies* 13.1) both provide historical discussion of the Australian context. I have here left out those texts that deal exclusively or mostly with gay men (such as James Keller's *Queer (Un)Friendly Film and*

Television), as there are quite different traditions, and this is not directly pertinent to my discussion in this monograph.
18. Although gossip channels allege that Aguliera is indeed bisexual.
19. Several analyses of this character and the manner in which lesbianism is represented in *Heartbeat* have been undertaken, most notably Sasha Torres's "Television/Feminism: *Heartbeat* and Prime Time Lesbianism," Darlene Hantzis and Valerie Lehr's "Whose Desire? Lesbian (Non)Sexuality and Television's Perpetuation of Hetereo/Sexism," and Marguerite Moritz's "Old Strategies for New Texts: How American Television Is Creating and Treating Lesbian Characters."
20. Didi Herman argues in "'*Bad Girls* Changed My Life": Homonormativity in a Women's Prison Drama" that *Bad Girls* "disrupts the WIP [women-in-prison] genre significantly. *BG* foregrounds lesbian heroines who have happy endings, and the normalization of lesbianism occurs outside as well as within the prison" (143). Herman also argues that the series, like *Queer as Folk*, "construct[s] an overt 'insider's view,' a homonormative space where lesbian and gay sexuality is both unremarkable and, potentially, desirable" (143).
21. Frameline awarded Channel 4 with a 1997 Frameline award for "outstanding contribution to gay and lesbian media arts" ("Tribute to Channel Four"). Channel 4 has since created and broadcasted the original *Queer as Folk* (1999–2000).
22. Iyari Limon has since come out as bisexual (see Kregloe).
23. This is in contrast to the sexually coded "magic" scenes, wherein Willow and Tara magically "deflower" a rose or "levitate" one another, for example. For further analysis of this issue, refer to Winslade, "Teen Witches, Wiccans, and Wanna-Blessed-Be's," or Beirne, "Queering the Slayer-Text."
24. J. Lawton Winslade addresses the connections between lesbian sex and magic in the series, Farah Mendlesohn discusses the denial of a queer element in Buffy and Willow's relationship, and Jes Battis discusses the hybridity of Willow's character, while Edwina Bartlem's "Coming Out on a Hellmouth" and my own "Queering the Slayer-text" specifically address Willow's lesbianism and the representation of queer sexuality by the series.
25. Examples include analyses of *Queer as Folk* (U.K.) by Munt, Johnson, and Alderson and analyses of *Queer as Folk* (U.S.) by Emig, Rasmussen, and Kenway, which all focus fairly exclusively on the portrayal of male homosexuality. The recent publication of *The New Queer Aesthetic on Television: Essays on Recent Programming*, edited by James R. Keller and Leslie Straatyner, contributed two further essays on *Queer as Folk* (see Peeren and Beirne), both of which discuss the construction and resignification of "queer" in the series, and Glyn Davis has recently published a monograph on *Queer as Folk*. The few analyses of female homosexuality in the series of which I am aware consist of a conference paper by Kate Monteiro and Sharon Bowers on lesbian representation in the American version and a more recent essay by Noble.
26. For further discussion of this adaptation of *Oranges Are Not the Only Fruit*, see Hilary Hinds's "*Oranges Are Not The Only Fruit*: Reaching Audiences Other Lesbian Texts Cannot Reach"; Marshment and Hallam's "From String of Knots to

Orange Box: Lesbianism on Prime Time"; and "Framing Experience: Case Studies in the Reception of *Oranges Are Not the Only Fruit*."

27. Erika Suderberg has discussed queer youth on the televison documentaries *An American Family* and *The Ride* as well as MTV's reality television series *The Real World*. For a summary of lesbian and bisexual women on American reality television up to 2004, see Warn's "Lesbians Come Out of the Reality TV Closet in 2004."

28. Both women mentioned their lesbian relationships in their bios on the official *Survivor* site, and the media frequently cited them as such. Ami claims to have come out the second day, while Scout came out around a week later. See "Survivor's First Axed Lesbian Tells All" (Behrens).

29. "There was a time when Australian television led the world in its representation of gays and lesbians in popular television" (Howes 39).

30. Perhaps some responsibility for such insistence on definitively locating the impact of each representation rests with "media watchdogs" such as Gay and Lesbian Alliance Against Defamation (GLAAD).

31. This is seen on television, in mainstream media, and perhaps most conspicuously in pornography. For more information on the manner in which gay and lesbian organizations such as the Human Rights Campaign have attempted to promote feminized images of lesbianism, see Phelan 1998.

32. For further reading on the relationship between lesbian and commodity cultures, see Danae Clark's "Commodity Lesbianism" (1991) and Alexandra Chasin's *Selling Out: The Gay and Lesbian Movement Goes to Market* (2000). I will later discuss *Queer as Folk*'s very positive attitude toward commodification and the more ambiguous perspective voiced by various characters in *Dykes to Watch Out For*.

33. Not having lived through the era myself, my understandings thereof have been reached through a fairly exhaustive study of both published and unpublished materials too numerous to list here. Authors that may be of particular interest to those interested in further investigation of the subject include Wilton, Healey, Duggan and Hunter, Martindale, Jeffreys, and Richardson. When engaged with this work, it is important to be mindful of the intensity of feeling generated by the debate, which has often resulted in distortion and conflation within individual accounts.

34. An advertising pamphlet for the latter, found in the Lesbian Herstory Archives, in an effort to differentiate itself from heterosexual pornography, promises, "sixteen speaking parts, twenty two dancing extras, original lesbian country music, horses, beautiful scenery, and exuberant energy! We have not styled our video after the usual porn market which is seventy to eighty percent sex. 'Hay Fever' offers plenty of sexual activity (around 38%) with a humorous plot, fun, and music" ("Tigress Presents").

35. There are various articles on lesbian performance art that involves stripping, transsexual striptease, and the role of the spectator in heterosexual striptease (which footnotes some lesbian striptease performances)—a great many of which have, like Dolan's article, been published in *The Drama Review*—but this

article remains the only one I am aware of that specifically discusses a lesbian strip show.

36. The use of the term "comix" arose in reaction to the very restrictive "Comics Code," which sought to sanitize comics and render them to be merely the province of child readers. The use of the term "comix" was intended to signify "alternative" comics intended for a more sophisticated, adult audience. According to Thalheimer: "[t]he term 'comix' . . . evolved to mark a distance between mass-market 'comics' (comic books approved by the Comics Code and comic strips homogenized through syndication) and underground 'comix' (works defying the Comics Code as well as a system seemingly more interested in profits than artistic integrity)" (59).

37. Sources conflict as to whether *Come Out Comix* was published in 1973 or 1974 and whether "A Modern Romance" was published in 1974 or 1975. As I have not been able to acquire a copy of either text, I cannot confirm which date is correct. Tuula Raikas also notes, "There is some controversy over who was the first lesbian cartoonist" (21).

38. The title changed to *Gay Comics* circa 1991 (Gay League, "LGBT Comics Timeline").

39. The 'zine became a popular form of publication in the 1990s. 'zines consist of small, self-published (usually photocopied) volumes and were particularly popular with both young Riot Grrrl–inspired feminists and the producers of comics. Comstock and Fraser provide specific description and analysis of 'zine culture and third-wave feminism.

40. For a fuller listing of lesbian and bisexual creators of the early 1990s, refer to Roz Warren's *Dyke Strippers: Lesbian Cartoonists A to Z*.

41. "The comics code is a set of regulatory guidelines primarily concerned with sex, violence, and language drawn up by publishers and enforced by the 'code authority,' a euphemism for the censor employed by the publishers. . . . [C]omics carrying the Comics Code Seal be ones that a parent can purchase with confidence that the contents uphold basic American moral and cultural values" (Nyberg vii, 175).

42. The code also stipulates that "[i]f it is dramatically appropriate for one character to demean another because of his or her sex, ethnicity, religion, sexual preference, political orientation, socioeconomic status, or disabilities, the demeaning words or actions will be clearly shown to be wrong or ignorant in the course of the story" (Nyberg 178).

43. All examples of these similarities are too numerous to list here; however, I will give some examples for illustrative purposes. Both Clarice and Melanie are lawyers for public causes, whereas their more femme partners Toni and Lindsay are more maternal and each have fervent desires for motherhood, each giving birth to a son. Both have a gay male best friend whom their partner ardently dislikes. When each goes back to work after a period as stay-at-home mothers, there is conflict between the partners as to whose commitments are the most important. Both Clarice and Melanie have one-night stands with other women in direct relation to issues they are experiencing created by their partner's (desires for)

motherhood. Both couples experience multiple bouts of sexual decline/disinterest in their relationships that forms the subject of various of their story lines. Toni and Lindsay later both cheat on their spouses after a period of restraining their desires, but while Toni does so with long-term friend and crush Gloria, Lindsay does so with male artist Sam, as will be discussed further in Chapter 2. The story line for Melanie and Lindsay's second child appears to be drawn from another couple in *Dykes to Watch Out For*'s childbearing experience—Sparrow and Stuart. Both Sparrow and Melanie are initially quite reluctant to become mothers, work too hard throughout their pregnancy (in Sparrow's case in her position as director at a women's shelter, Melanie on a custody case for lesbian mothers), which induces premature contractions, and are confined to total bed rest for the remainder of their pregnancy, where they each become extremely irritated by their respective partners. Each then has a girl, Sparrow's named Jiao Raizel (known as JR) and Melanie's named Jenny Rebecca (also known as JR). Significantly, in each parallel, *Queer as Folk*'s characters are whitened. Latina Toni and African American Clarice become Caucasian Lindsay and Jewish Melanie, while multiracial Jiao Raizel (Chinese and Hebrew names respectively to reflect her ancestry) becomes Jenny Rebecca. *The L Word* is also not immune from similarities with *Dykes to Watch Out For*. Bechdel's character Toni makes the statement during an argument about potential sperm donors for her and Clarice's child that "Well, I *know* we decided on a Latino donor. The kid's gonna have a hard enough time with an interracial lesbian couple for parents, let alone being mixed-race herself" (*Dykes to Watch Out For: The Sequel* 99); when *The L Word*'s Bette proposes an African American donor to her wife Tina, Tina responds "don't you think, on top of everything else, to also have two moms . . . that is a *lot* of otherness to put on one child" (1.1).

CHAPTER 3

1. While such characters as *Buffy the Vampire Slayer*'s Willow may have received proportionately more screen time, for example, little attention was paid to her lesbianism.
2. This aspect is not necessarily a gesture of inclusion but may simply be due to more ensemble focus and, as was mentioned earlier, greater screen time of the Showtime series (while the original series included only ten episodes, the American series had eighty-three episodes).
3. There is one reference in 1.16 to Lindsay becoming a "scary political dyke" in college and circulating a petition to bring in castration as a punishment for date rape (which may have been in jest), and she is somewhat involved in various committees and fundraisers for the gay and lesbian center; however, in these latter instances, she is depicted as primarily doing so for Melanie's sake, who takes a more leading role.
4. This is despite their infidelities. Their dishonest monogamy (also seen in other single-episode characters depicted as assimilationist) is posed against the "honest nonmonogamy" of the series' queer-coded characters.

5. Examples of such desexualised depictions include films such as *Serving in Silence: The Margarethe Cammermeyer Story* (1995), *Boys on the Side* (1995) and *My Best Friend's Wedding* (1997) or television shows such as *Ellen* (1994–1998) or *Will & Grace* (1998–2006).
6. I use this term not in the Marxist sense but rather because the hierarchy generated is done so as a reaction to both real and perceived threats.
7. It is not made clear whether she is here discussing the British or American versions of the series because neither is referenced or discussed beyond this sentence. Considering the geographic and temporal location of the author, one could make an educated guess that this statement is primarily made in reference to the first and second seasons of the U.S. version of the program.
8. Debbie, a woman who is the epitome of what could be known as the "queer straight" (PFLAG member, constantly emblazoned with rainbows, and a part of the gay community through her brother, son, and work in the Liberty Diner), is strangely granted access to the world of the homosexual through the singing of the song "for she's a jolly good homo" in 5.3. This is despite lesbians being linguistically denied the classification of "homo" by a statement in the same season: "it's the seating chart for our wedding dinner . . . homo, lesbian, homo, lesbian, homo, lesbian" (5.12).
9. I find it unlikely that Russell T. Davies (writer of the original series) was unaware of this latter usage and used it to his advantage; however, this of course is also possible, and the resultant usage is a coincidence. The differences created by the switch from queer as descriptive of a noun to queer as descriptive of a verb should not be underestimated.
10. Although interchangeable, there are subtle distinctions in meaning between the terms dyke and lesbian. The reclaimed understanding of "dyke" as it emerged during and after the sex wars is that it is both deeply connected with sex and with the claiming of despised identity and practices. This usage is a reaction to some definitions of "lesbian" that sought to universalize the term to apply to any "woman identified woman" regardless of sexual orientation. Despite these histories, in contemporary colloquial usage, "dyke" instead of "lesbian" is, like queer, seen as a more open term, and some women who are not exclusively homosexual identify with it. The more polite "gay women" has become increasingly popular, particularly among conservative homosexual women, the most obvious television examples being Ellen DeGeneres or Ivana on *Big Brother*, as is discussed in Chapter 1.
11. Vera Wang is a high-profile fashion designer who specializes in wedding dresses.
12. See, for example, William Eskridge and Andrew Sullivan.
13. The whole quotation is: "I think we've said everything we wanted to say, about HIV and AIDS; the crystal meth addictions; discrimination; a political climate that's becoming far more conservative and oppressive; gay parenthood; the conflict in the community between the assimilationists and those who want to continue a queer lifestyle, whom Brian represents. I think there's a huge conflict between those two elements right now." Ron Cowen was being interviewed at the wrap of the fifth and final season of *Queer as Folk*.

14. See Beirne, "Embattled Sex," for further discussion of *Queer as Folk*'s discussion of, and positioning toward, assimilationist politics.
15. See, for example, Emmett's "Am I to believe I'm actually getting decorator tips from l-lesbians?" (which Melanie and Lindsay succeed in finding amusing) and the following exchange:

> *Emmett:* Can you believe those dykes telling us how we should decorate our house? Don't they realise Laura Ashley went out with rotary phones?
> *Ted:* Now now, show a little compassion, I mean they can't help it if they're design-challenged. (3.6)

16. Although Carter notes that "[t]he existence of this lesbian and her supposed role in the Stonewall Riots have always been the most controversial aspects of the riots, with some prominent commentators displaying skepticism about her," he goes on to assemble an impressive array of witnesses and evidence that attests to the central role played by this woman in beginning the resistance at Stonewall. Indeed, as one anonymous letter recounts, "[e]verything went along fairly peacefully until . . . a dyke . . . lost her mind in the streets of the West Village—kicking, cursing, screaming and fighting" (Carter 151). Another witness, Kevin Dunn, "who had seen a fair percentage of the gay men leaving the Stonewall Inn camping and posing, noticed that 'the lesbians who came out were not in a good humour to do a little pose. They were resistant about being busted'" (153), while Philip Eagles notes that "the cops were roughing up some of the lesbians. . . . Of course, the butch lesbians were among the first to start fighting back as I remember . . . and so they were getting beaten and hit" (153) Tree, bartender at the Stonewall Inn and a witness to the Stonewall riots in 1969, stated that the name of the lesbian who threw the "first punch" at Stonewall was butch lesbian Stormé DeLaverie (during a conversation with Tree in New York on 26 October 2004). Stormé DeLaverie is the subject of Michelle Parkerson's documentary film *Stormé: Lady of the Jewel Box* (1987; thank you to Patricia White for this reference).
17. See further discussion thereof in Beirne, "Embattled Sex," 49–51.
18. For an analysis of the impacts of commodifying homosexuality, refer to Alexandra Chasin's *Selling Out: The Gay and Lesbian Movement Goes to Market* or pages 235–72 of Walters' *All the Rage*.
19. U.S. television has a long tradition of broadcasting same-sex kisses between women during "sweeps week" in order to boost its ratings to entice advertisers. See Virginia Heffernan's *New York Times* article, "It's February; Pucker Up, TV Actresses."
20. See Healey (especially "Of Quims and Queers," 181–202) or any contemporary lesbian magazine.
21. I am not agreeing with the posing of "sex-positive" and feminist against one another; I am just pointing out a popular narrative commonly set up in lesbian-specific as well as mainstream publications and discourses.

22. The terminology I am using here comes from Seidman's distinction between good and bad sexual citizens as outlined in *Beyond the Closet*, 150–59.
23. In the episode in which Melanie and Lindsay get married, for example, Debbie tells Justin that his place is at the wedding "*with [his] family*" instead of at the white party with Brian. Melanie's wedding vows also echo this feeling: "thanks to our friends, or should I say *our family*, not even the stars or the planets could keep us from exchanging our vows" (2.11). This vision of friends becoming family is, of course, not unusual; one can witness sentimental statements about "friends becoming family" in almost any mainstream sitcom.
24. However, it could be argued that having two parents "who love each other" indeed mirrors traditional understandings of family.
25. "A beard is a woman or man who disguises the sexual interest of her or his partner" (Hardie 276).
26. An example of such a commentator is William Eskridge, whose monograph *The Case for Same-Sex Marriage: From Sexual Liberty to Civilized Commitment* asserts that "[s]ince 1981 and probably earlier, gays were civilizing themselves. Part of our self-civilization has been an insistence on the right to marry" (58). Further examples of such arguments can be found in Warner's discussion of gay marriage on pages 81–148 of *The Trouble with Normal*.
27. This reinforces the popular notion that lesbians "turn to muff" not because they like it but because of male chauvinism.

CHAPTER 4

1. For general examples of arguments regarding feminized lesbian images in the mainstream, see Ciasullo's analysis of 1990s mainstream images or Inness's discussion of 1990s magazine images of lesbianism. More specific debates over *The L Word* include Lo's "It's All about the Hair" and Warn's counterpoint "Too Much Otherness," as well as Nyong'o's "Queer TV: A Comment."
2. For discussion of the latter formulation from a femme perspective, see Lisa Walker's account of Judith Butler's position on the desired butch in "How to Recognize a Lesbian," 885.
3. I am not suggesting that such a phenomenon is new or unique to the series. These practices are very common to the growing number of depathologized as well as pathologized images of lesbians on television and in film.
4. This encounter is used to deride not only the proponents of identity politics but also political engagement with issues of racism, sexism, and capitalism and with politically directed anger more generally. Bette reads a small excerpt from Yolanda's collection *Sistah, Stand Up* (perhaps a reference to Lorde's *Sister Outsider*, particularly as Yolanda is both a poet and a critic): "On being a black, socialist, feminist lesbian, working to overthrow the white, male, capitalist patriarchy." Bette prefaces her reading with the comment on the collection's title, "with an 'h,' might I add," and then goes on to comment, "I noticed 'lesbian' comes last."
5. Ilaria Urbinati, "owner of the hip L.A. show boutique Satine, joined the *The L Word* team to revamp its image for the second season. . . . Looks like the other L

word is luxe" ("*L Word* to fill SATC Style Gap?" 2004). In the "*L Word* Fashion Extra" in the first season DVD box set, Urbinati introduces the fashion for the second season, stating that "Showtime wanted to really take the fashion of the show to the next level," and notes that the vision of the new fashion consultants for Dana's character is "a kind of country-clubbish vibe."

6. Inness discusses "how women's magazines operate to 'normalize' lesbians by assuring heterosexual readers that lesbians are, indeed, very much like heterosexuals, partially stripping lesbians of their identities" (53), and this is often done by emphasizing the physical attractiveness of the magazine's interview subjects.
7. The most marked of these came from Kelly Lynch (Ivan), who announced during an interview in *Curve* that "half of the cast is gay, another third of them are bisexual, another couple of them maybe are confused about who they are but maybe, you know, have some issues" (Anderson-Minshall 38–39).
8. Douglas also notes that "Tina's character shows us the problems of lesbian couples emulating heteronormativity especially because Bette treats Tina as a traditional housewife and as an object" ("That's Not Me" 55).
9. During the first episode, the usually mild-mannered and physically/emotionally weak Tina flips over a large table in a coffeeshop in anger at Bette. She then proceeds throughout the season to "hold all the cards" in terms of her and Bette's relationship (2.6). Although, or perhaps because of, being very pregnant, she manages to gain funding from and then seduce a wealthy philanthropist (Helena), and at one point, she is having sex with both Helena and Bette. After having sex with Tina once again, Bette confesses to her therapist that it was also a saddening experience because "we don't . . . well, like, she doesn't belong to me anymore. . . . And . . . and she did things that we had never done together. And it was like she was so . . . [sighs] free. . . . I mean, I always treated her so gingerly, you know, like she was some fragile thing. And now, even though she's pregnant, it's like, it's like she's . . . it's like she's unbreakable" (2.9).
10. Shane's parenting of half-brother Shay in season four is a possible counter to this, as it leads her to a relationship and "instant family" with single mother Paige (Kristanna Loken), though this is clearly not directly "motherhood."
11. Although, as is noted by C. Taylor "even if racially, Ion Overman is difficult to read: she has been claimed and championed by bi-racial/multiracial, Hispanic and African-American lesbians alike."
12. This is also exercised via actual mirrors in the text. For example, in the first episode we see Marina and Jenny in the bathroom, each reflected in two separate mirrors with individual large gold frames. Jenny asserts that she would like to see Marina again (this is before they first have sex), and Marina symbolically pulls her into the frame of her mirror.
13. That Marina's love for Jenny destroys Marina's relationship and business and causes her to attempt to take her own life (2.1)—which, while it doesn't actually kill her, is the device of her removal from the series (effectively "killing" her by making her cease to exist)—furthers the allusion to *Dracula* and many pulp lesbian novels.

14. Whereas in this literature, it is Eastern Europe in relation to Britain that is posited as the "other," a polyphonous source of fear and desire, in the case of Marina in *The L Word*, it is Western Europe (we see Marina speaking flawless French and Italian) in relation to America, with Europe's associations with freer attitudes toward sexuality, that is the site of cultural anxiety.
15. During a fashion shoot and interview for *Vogue*, for example, Moenning is quoted as saying, "I'm a total tomboy at heart. Incorporating menswear into my wardrobe is something that just happens when I dress myself" (Arakas 408).
16. See, for example, Guy Trebay's "The Secret Power of Lesbian Style" and the general fashionista adoration of lesbian *Sex and the City* stylist Patricia Field.
17. The instances of fierce desire in *The L Word* are generally extramarital, bound to fail, and coexistent with a fierce unhappiness. The overtly more successful marriage-type relationships in the series are, by contrast, propelled by a less violent and grief-stricken desire but are seen as boring and stifling to the selfhood of their participants. Both forms of couplings are portrayed as strangely hollow and are characterized by lack.
18. The first time the audience sees Shane being directly touched/topped is as a result of a game with Carmen, and she seems fairly uncomfortable with this.
19. The show's attitude toward queering gender cannot be properly analyzed without reference to its first presentation of an ambiguously gendered/sexually identified person in the series, Lisa the "lesbian identified man," whom Alice, an out and proud bisexual, briefly dates. Lisa's identity as a lesbian appears to be primarily expressed in the series through an attachment to Reiki, vegetarianism, and an extreme clinginess, which apparently cause him to "do lesbian better than any lesbian [she] knows" (Alice in 1.9). Male, self-identified lesbians have also been a subject under discussion in the lesbian community, particularly in the early 1990s, as is seen in Jacquelyn N. Zita's "Male Lesbians and the Postmodernist Body." Like Zita's work, then, this story line could be read as a critique of postmodernist excesses in bodily dislocation. However, due to lack of the contemporary manifestations of this debate and the ongoing debates about transsexual women in lesbian cultural spaces, this could also be metaphorically read as a strange, hyperbolic, and insensitive displacement of the debates about transsexual women who are lesbians and their inclusion in lesbian communities. The characters all react very negatively and rudely to Lisa, except Kit, who asserts that "[i]f the dude wanna give up his white man rights to be a second-class citizen, then hey, welcome to our world" (1.6). Jennifer Moorman discusses this story line in "'Shades of Grey': Articulations of Bisexuality in *The L Word*."
20. At a seminar in May 2004, Chaiken stated that the character was inspired by "a lesbian 'sexpert' named Ivan . . . from Vancouver" (Chaiken paraphrased by Kristen) who ran a seminar on lesbian sex for the (primarily heterosexual) actors before commencement of shooting. Although not anywhere listed as a "sexpert," as far as I can tell, this has to be Coyote, as the intersections between their names, gender definitions, and dress and Coyote's residence in Vancouver seem too strong for this to be otherwise.

21. With the exception of the very first episode, many episodes of *The L Word* begin with a scene that is not immediately or obviously linked to the narrative. During the first season, for example, a character, idea, or object from these scenes later becomes linked to the narrative in either a direct or obscure way.
22. "In the demographic calculus that lay behind the decision to underwrite the series, Gary Levine, Showtime's vice president for original programming, told the *New York Daily Times* its potential appeal to non-lesbian viewers rested on the understanding that 'lesbian sex, girl-on-girl, is a whole cottage industry for heterosexual men'" (Sedgwick B10).
23. A great deal of the marketing of *The L Word* appears to be toward the heterosexual population—the marketing appears (probably rightly) to operate on the presumption that lesbians will watch regardless.
24. The insertion of such a perspective mirrors that of "Scott Seomin, the entertainment media director of the Gay and Lesbian Alliance Against Defamation, [who] sees the porn connection as a smart crossover move. 'If they pull them in and get them hooked on the titillation factor,' he told the *Daily News*, 'that straight male is going to learn about the lives of lesbians'" (Sedgwick B10).

Chapter 5

1. For more information on the Butler decision, see Brenda Cossman et al *Bad Attitude/s on Trial: Pornography, Feminism and the Butler Decision*.
2. To Bensinger, the two are not only different but also not really related; rather, they are pitted throughout the article as permanently at odds with one another.
3. Having always included transsexual women in its definition of "women-only," *Gurlesque* has now changed the "women-only" label on their promotional material to "for women and trannies."
4. Author's interview with Sex Intents and Glita Supernova, 17 Jan. 2005. Nonverbal articulations have been removed from this and all subsequent quotations for clarity and ease of reading.
5. Here Dolan is referring to the *On Our Backs*–produced *BurLEZk*, which took place at "Baybrick's, a now-defunct lesbian bar in San Francisco" (63).
6. Although in her analysis Liepe-Levinson focuses principally on both male and female heterosexual shows, she occasionally uses the gay, lesbian, and transsexual strip events she also attended as a point of comparison. Of the lesbian events, she writes in a footnote: "Lesbian strip events, like many of those for straight women, tend to be subset within other events and entertainments. . . . In addition to the party circuits and ad hoc performances in bars and clubs, strip events for lesbians researched in the initial stages of the study included New York City's Brooklyn dance club; Spectrum Disco, which featured the performances of one male and one female stripper on weekend nights for mixed audiences; Tracks, a gay nightclub in Washington DC, that presented a weekly event called "Lesbo-A-Go-Go," an exotic dance show for lesbians that included stripping but no actual nudity; . . . and Debi Sundahl's *Burlezk* show, which originated at the Baybrick in San Francisco and made a number of road tours" (34).

7. Examples include bisexual comic book artist Colleen Coover's ongoing comic *Small Favours* (2002–) and Petra Waldron and Jennifer Finch's *The Adventures of a Lesbian College Schoolgirl* (1998).
8. I attended sixteen lesbian-themed films at the festival. The only other sessions that had a full or almost full cinema were *Goldfish Memory*, which contained a mixture of lesbian, gay, bisexual, and straight characters. In contrast, the gay-male themed films were characterized by queues that filled the lobby well before the beginning of each film. Another popular (though not quite as popular) session was *Beauteous Bingebabes*, a collection of lesbian-themed short films that allowed entrance to those over fifteen years old instead of the usual over-eighteen policy of the festival.
9. It should be noted that the description of *Mango Kiss* as an S/M comedy is not entirely accurate in terms of both its content and its attitudes towards sex-wars era culture.
10. Kings Cross is Sydney's red light district.
11. The hidden nature of female ejaculation is narratively portrayed in *The L Word*'s "Lies, Lies, Lies" (1.4) through the character Dana, who expresses intense shame at having ejaculated due to her ignorance of what exactly had occurred.
12. Nonverbal articulations have been edited out of this transcript for clarity.
13. Mardi Gras is the equivalent of North American Pride Festivities, "a three-week festival of gay and lesbian culture—arts, sport, debate—held in Sydney, Australia during February, which culminates in a Parade. . . . The Festival is the biggest of its kind in the world" (Wherrett 1).
14. It should be noted that this was paired with a wig with long braids and fishnets—rendering her more feminine than her usual garb—perhaps explicitly done to undermine the signification of the dildo.
15. The very use of the term "patriarchal" on the web site (Gurlesque, "About Us"; also used at several points during the interview), rather than a more contemporary term perceived to be more "queer," is itself interesting.

Chapter 6

1. I was born the year the sex wars ostensibly began (1982) and came out 1998–1999.
2. This *understanding* of previous generations that can be given by comics is an important one. Before propornography academics and practitioners condemn those feminists who engaged in antipornography activism in the early mid-1980s as evil and antilesbian, for example, they would do well to read the "Anti-Porn Fanatic" issues of *Liliane, Bi-Dyke* (Nov. 1993/Oct. 1994) or Bechdel's "Sentimental Education" (*Unnatural Dykes* 111–42)—not to change their positions, but to attempt to gain an understanding of the sorts of reasons women engaged in the other side of these politics and indeed often did so before shifting to a sex-radical perspective.
3. This is asserted by Sparrow, the only heterosexually conceiving woman in the narrative, upon asking partner Stuart to have a vasectomy and "make some

deposits at a sperm bank first. Then we can inseminate like normal people when we're ready" (*Dykes and Sundry Other* 104).
4. Citations to particular *Dykes to Watch Out For* comic strips include their strip number to allow online readers to easily locate them, as well as name and page number of the volume they were (re)published in.
5. All *Dykes to Watch Out For* text is written in uppercase letters with bold text used to add emphasis. To allow both ease of reading and consistency with the other quotes throughout my text, I have altered these quotes into a grammatically consistent combination of uppercase and lowercase text with italics instead of bold for emphasis.
6. Her perspective also prefigures Halberstam's in "The I Word."
7. For further discussion on the financial dimension of mainstreaming, see Alexandra Chasin's *Selling Out: The Gay and Lesbian Movement Goes to Market*.

Works Cited

Addiego, Frank, Edwin G. Belzer Jr., Jill Comolli, William Moger, John D. Perry, and Beverly Whipple. "Female Ejaculation: A Case Study." *The Journal of Sex Research* 17.1 (1981): 13–21.
Alderson, David. "Queer Cosmopolitanism: Place, Politics, Citizenship, and *Queer as Folk*." *New Formations: A Journal of Culture/Theory/Politics* 55 (2005): 73–88.
All My Children. Created by Agnes Nixon. ABC, 1970–.
All Over Me. Dir. Alex Sichel. Perf. Alison Folland, Tara Subkoff, Wilson Cruz, and Leisha Hailey. Sichel Sisters, 1997.
Ally McBeal. Created by David E. Kelley. Fox, 1997–2002.
America's Next Top Model. Created by Tyra Banks. UPN, 2003–.
Andelman, Bob. "Mr. Media: Come Out, Come Out, Wherever You Are." *Andelman.com*. 2 Dec. 1996. 13 Aug. 2005. <http://www.andelman.com/mrmedia/96/12.02.96.html>.
Anderson-Minshall, Diane. "L Is for Lynch" [Interview]. *Curve* 14.6 (Oct. 2004): 38–40. Expanded Academic ASAP. Gale Group/InfoTrac. U of Sydney Libraries. 7 Sep. 2005 <http://infotrac.galegroup.com/itweb/>.
Angels in America [miniseries]. Dir. Mike Nichols. Written by Tony Kushner. HBO, 2003.
Arakas, Irini. "What Women Want." *Vogue Magazine* Sept. 2004: 400–408.
A Question of Love [telemovie]. Dir. Jerry Thorpe. Perf. Gena Rowlands and Jane Alexander. ABC, 1978.
Bad Attitude: A Lesbian Sex Magazine. Boston: Phantasia Publications, 1984–1987.
Bad Girls. Created by Maureen Chadwick, Eileen Gallagher, and Ann McManus. ITV1. 1999–.
Bakhtin, Mikhail. "The Grotesque Image of the Body and Its Sources." *Rabelais and His World*. Trans. Helene Iswolsky. Cambridge and London: The M.I.T. Press, 1968. 303–67.

Works Cited

Barthes, Roland. "Striptease." *Mythologies.* Trans. Annette Levers. 1972. London: Jonathan Cape Ltd., 1974. 84–87.

Bartlem, Edwina. "Coming Out on a Hell Mouth." *Refractory: A Journal of Entertainment Media* 2 (2003). 12 Jul. 2003. <http://www.refractory.unimelb.edu.au/journalissues/vol2/edwinabartlem.htm>.

Battis, Jes. "'She's Not All Grown Yet': Willow as Hybrid/Hero in *Buffy the Vampire Slayer.*" *Slayage: The Online International Journal of Buffy Studies* 8 (2003). 10 Jul 2003. <http://www.slayage.tv/essays/slayage8/Battis.htm>.

Bawer, Bruce. *A Place at the Table: The Gay Individual in American Society.* 1993. New York: Touchstone, 1994.

Beadling, Laura. "Queerness and the Lesbian-Identified Man in *The L-Word.*" Conference paper. Midwest Modern Language Association annual conference, St. Louis, November 2004. 4 Jul. 2005.

Bechdel, Alison. *Dykes and Sundry Other Carbon-Based Life-Forms to Watch Out For.* Los Angeles: Alyson Publications, 2003.

———. "Dyke March." *DTWOF: The Blog.* 12 Jun. 2005. 18 Aug. 2005. <http://alisonbechdel.blogspot.com/2005/06/dyke-march_12.html>.

———. *Dykes to Watch Out For.* Ann Arbor: Firebrand Books, 1986.

———. *Dykes to Watch Out For: The Sequel.* Ithaca: Firebrand Books, 1992.

———. *Hot, Throbbing Dykes to Watch Out For.* Ithaca: Firebrand Books, 1997.

———. *The Indelible Alison Bechdel: Confessions, Comix, and Miscellaneous Dykes to Watch Out For.* Ithaca: Firebrand Books, 1998.

———. "In the Good Old Summertime." *Planet Out Comics* #471. 24 Aug. 2005. 1 Dec. 2005 <http://www.planetout.com/entertainment/comics/dtwof/archive/471.html>.

———. *Invasion of the Dykes to Watch Out For.* Los Angeles: Alyson Publications, 2005.

———. *Post Dykes to Watch Out For.* Ann Arbor: Firebrand Books, 2000.

———. *Spawn of Dykes to Watch Out For.* Ithaca: Firebrand Books, 1993.

———. "The Temptations." *Planet Out Comics* #467. 1 Jun. 2005. 1 Dec. 2005 <http://www.planetout.com/entertainment/comics/dtwof/archive/467.html>.

———. *Unnatural Dykes to Watch Out For.* Ithaca: Firebrand Books, 1995.

Behrens, Web. "Survivor's First Axed Lesbian Tells All." *Advocate.com.* 7 Dec. 2004. 3 Feb. 2005. <http://www.advocate.com/html/stories/929/929_survivor.asp>.

Beirne, Rebecca. "Embattled Sex: Rise of the Right and Victory of the Queer in *Queer as Folk.*" *The New Queer Aesthetic on Television: Essays on Recent Programming.* Ed. James R. Keller and Leslie Stratyner. Jefferson, NC: McFarland, 2006. 43–58.

———. "Introduction: A History of Lesbian Television Criticism." *Televising Queer Women: A Reader.* Ed. Rebecca Beirne. New York: Palgrave MacMillan, 2007. 1–15.

———. "Lipstick and Lesbians: Visibility in *The L Word.*" Conference paper. Midwest Modern Language Association annual conference, St. Louis, November 2004. 4 Jul. 2005.

———. "Queering the Slayer-Text: Reading Possibilities in *Buffy the Vampire Slayer.*" *Refractory: A Journal of Entertainment Media* 5 (2004). 10 Oct. 2004. <http://www.refractory.unimelb.edu.au/journalissues/vol5/beirne.htm>.

———. "Screening the Dykes of Oz: Lesbian Representation on Australian Television." *Journal of Lesbian Studies* 13.1, forthcoming 2009.

Becker, Ron. *Gay TV and Straight America*. Piscataway, NJ: Rutgers UP, 2006.

Belzer, Edwin G. Jnr. "Orgasmic Expulsions of Women: A Review and Heuristic Inquiry." *The Journal of Sex Research* 17.1 (1981): 1–12.

Bensinger, Terralee. "Lesbian Pornography: The Re/Making of a Community." *Discourse* 15.1 (1992): 69–93.

Better than Chocolate. Dir. Anne Wheeler. Perf. Wendy Crewson, Karen Dwyer, and Peter Outerbridge. Rave Film Inc., 1999.

Betty. "The L Word Theme." *The L Word: Season 2 [Soundtrack]*. Tommy Boy, 2005.

Big Brother. Created by Marc Pos. Endemol, 1999–.

Bikini Kill. "Rebel Girl." *The C.D. Version of the First Two Records*. Kill Rock Stars, 1994.

Blackman, Inge, and Perry, Kathryn. "Skirting the Issue: Lesbian Fashion for the 1990s." *Feminist Review* 34 (1990): 67–78.

Bolonik, Kera. *The L Word: Welcome to Our Planet*. New York: Fireside, 2006.

———. "Not Your Mother's Lesbians." *New York* 12 Jan. 2004. 19 Jan. 2004. <http://www.newyorkmetro.com/nymetro/news/features/n_9708/>.

Bornstein, Kate. *Gender Outlaw: On Men, Women and the Rest of Us*. New York and London: Routledge, 1994.

Boys on the Side. Dir. Herbert Ross. Perf. Whoopi Goldberg, Mary-Louise Parker, and Drew Barrymore. Alcor Films, 1995.

Braddock, Paige. *Jane's World* 1998–.

———. *Jane's World Vol. 1*. Sebastopol, CA: Girl Twirl Comics, 2003.

———. "Will Eisner Comic Industry Awards: *Jane's World* nominated for best humor book." *Paigebraddock.com*. 19 Apr. 2006. <http://www.paigebraddock.com/>.

Bradford, Judith, Kirsten Barrett, and Julie A. Honnold. "2000 Census and Same-Sex Households: A User's Guide." *National Gay and Lesbian Task Force*. 1 Oct. 2002. 23 Aug. 2005. <http://www.thetaskforce.org/reslibrary/census.cfm>.

Bright, Susie. "The First Time: Summer 1984." 1984. *Susie Sexpert's Lesbian Sex World*. Pittsburgh and San Francisco: Cleis Press, 1990. 17–20.

Brill, Stephanie A. *The Queer Parent's Primer: A Lesbian and Gay Families' Guide to Navigating Through a Straight World*. Oakland, CA: New Harbinger Publications, 2001.

Brookside. Created by Phil Redmond. Channel 4, 1982–2003.

Buffy the Vampire Slayer. Created by Joss Whedon. WB, 1997–2001; UPN, 2001–2003.

Burgess, Allison. "There's Something Queer Going On in Orange County: The Representation of Queer Women's Sexuality in *The O.C.*" *Televising Queer Women: A Reader*. Ed. Rebecca Beirne. New York: Palgrave Macmillan, 2007. 211–27.

But I'm a Cheerleader. Dir. Jamie Babbit. Perf. Natasha Lyonne, Clea DuVall, RuPaul Charles, and Cathy Moriarty. Ignite Entertainment, 1999.

Butler, Heather. "What Do You Call a Lesbian with Long Fingers? The Development of Dyke Pornography." *Porn Studies*. Ed. Linda Williams. Durham, NC: Duke UP, 2004. 167–97.

Butler, Judith. "Desire and Dread: The Meanings of What We Do" [Review of lesbian sex magazines]. *Gay Community News* 12.9 (15 Sept. 1984): 3, 7.

———. "Imitation and Gender Insubordination." *Inside/Out: Lesbian Theories, Gay Theories*. Ed. Diana Fuss. New York and London: Routledge, 1991. 13–31.

Byrd, Ronald. "Characters: Destiny and Mystique." *Gay League* Web site. 24 Aug. 2004. <http://www.gayleague.com/gay/characters/display.php?id=49>.

Camper, Jennifer. "*Cookie Jones, Lesbian Detective.*" *Gay Community News*, 1980.

———, ed. *Juicy Mother (Number One: Celebration)*. Brooklyn: Soft Skull Press, 2005.

———. *Rude Girls and Dangerous Women*. Bala Cynwyd, PA: Laugh Lines Press, 1994.

———. *SubGURLZ*. San Francisco: Cleis Press, 1999.

Capsuto, Steven. *Alternate Channels: The Uncensored Story of Gay and Lesbian Images on Radio and Television*. New York: Ballantine Books, 2000.

Carter, David. *Stonewall: The Riots that Sparked the Gay Revolution*. New York: St. Martin's Press, 2004.

Case, Sue-Ellen. "Toward a Butch-Feminist Retro-Future." *Cross-Purposes: Lesbians, Feminists, and the Limits of Alliance*. Ed. Dana Heller. Bloomington and Indianapolis: Indiana UP, 1997. 205–20.

———. "Towards a Butch-Femme Aesthetic." *Discourse* 11.1 (1988–89): 55–73.

Cave, Nick. "Into My Arms." *The Boatman's Call*. Warner Bros, 1997.

Chambers, Samuel A. "Heteronormativity and *The L Word*: From a Politics of Representation to a Politics of Norms." *Reading the L Word: Outing Contemporary Television*. Ed. Kim Akass and Janet McCabe. London: I.B. Tauris, 2006. 81–98.

Chasin, Alexandra. *Selling Out: The Gay and Lesbian Movement Goes to Market*. New York: Palgrave, 2000.

Ciasullo, Ann. "Making Her (In)Visible: Cultural Representations of Lesbianism and the Lesbian Body in the 1990s." *Feminist Studies* 27.3 (2001): 577–608.

Cheer Up!: Don't Let the System Get You Down [Documentary] [short documentary] Dir. Jen Nedbalsky and Mary Christmas. NYC Radical Cheerleaders, 2004.

Clark, Danae. "Commodity Lesbianism" *Camera Obscura* 25–26 (1991): 180–201.

Clunis, D. Merilee, and G. Dorsey Green. *The Lesbian Parenting Book: A Guide to Creating Families and Raising Children*. 1995. New York: Seal Press, 2003.

Cohen, Leonard. "I'm Your Man." *I'm Your Man*. Sony, 1988.

Cole, Susan G. "Toon Fine-tuned: *Post-Dykes to Watch Out For* by Alison Bechdel." *Now Toronto* 22–28 Jun. 2000. 6 Nov. 2005. <http://www.nowtoronto.com/issues/2000-06-22/book_reviews.html>.

Common Ground [telemovie]. Dir. Donna Deitch. Perf. Erik Knudsen, Brittany Murphy, and Jason Priestley. Showtime, 2000.

The Complex. Created by Jennifer Lane and Katherine Brooks. Online series, c. 2003. <http://www.thecomplex.tv/>.

Comstock, Michelle. "Grrrl Zine Networks: Re-Composing Spaces of Authority, Gender, and Culture." *JAC: A Journal of Composition Theory* 21.2 (2001): 383–409.
Conway, Mary T. "Inhabiting the Phallus: Reading *Safe Is Desire*." *Camera Obscura* 38 (1996): 133–61.
———. "Spectatorship in Lesbian Porn: The Woman's Woman's Film." *Wide Angle* 19.3 (1997): 91–111.
Coover, Colleen. *Small Favors*. 2001–.
———. *Small Favors Girly Porno Comic Collection: Book One*. Seattle: Eros Comix, 2002.
Cossman, Brenda, Shannon Bell, Lisa Gotell, and Becki L. Ross. *Bad Attitude/s on Trial: Pornography, Feminism, and the Butler Decision*. Toronto, Buffalo, and London: U of Toronto P, 1997.
Cotter, Catherine Louise. Co-ordinator, Media Committee, Lesbian Feminist Liberation Inc. Letter to television networks. New York, 11 Feb. 1975. Located in Lesbian Herstory Archives, Brooklyn, NY, Sept. 2004.
Cottingham, Laura. *Lesbians Are So Chic . . . That We Are Not Really Lesbians at All*. London and New York: Cassell, 1996.
Cowen, Ron, and Daniel Lipman. "Interview with Cast and Producers of *Queer as Folk* on *Larry King Live* 24 Apr. 2002, 21:00 ET." *Queer as Folk Addiction* Web site 5 Sep. 2003. <http://www.angelfire.com/home/qaf/king_1.html>.
The Crash Pad. Dir. Shine Louise Houston. Perf. Dylan Ryan, Roxie Ryder, Dusty Ryder, Jo, Jiz Lee, and Shawn. Pink and White Productions, 2005.
Cromwell, Jason. *Transmen and FTMs: Identities, Bodies, Genders, and Sexualities*. Champaign: U of Illinois P, 1999.
Croome, Rodney. "'This Is Where I Want to Live': Supporting Same Sex Attracted Rural Youth." Paper delivered at the third Health in Difference Conference, Adelaide, October, 1999. 22 Oct. 1999. Rpt. on Web site. 29 Aug. 2005. <http://www.rodneycroome.id.au/other_more?id=173_0_2_0_M14>.
Crowley, Vicki. "Drag Kings 'Down Under': An Archive and Introspective of a Few Aussie Blokes." *The Drag King Anthology*. Ed. Donna Troka, Kathleen LeBesco, and Jean Noble. Binghamton, NY: The Haworth Press, 2002. 285–308.
Curve [periodical]. San Francisco: Outspoken Enterprises Inc., 1996–.
Dangerous Living: Coming Out in the Developing World. Dir. John Scagliotti. Narrated by Janeane Garofalo. First Run Features, 2003.
Dark Angel. Created by James Cameron and Charles Eglee. Fox, 2000–2002.
Davis, Glyn. *Queer as Folk*. London: BFI Publishing, 2007.
D.E.B.S. Dir. Angela Robinson. Perf. Sara Foster, Jordana Brewster, Jill Ritchie, and Meagan Good. Destination Films, 2004.
de Lauretis, Teresa. "Fem/Les Scramble." *Cross-Purposes: Lesbians, Feminists, and the Limits of Alliance*. Ed. Dana Heller. Bloomington and Indianapolis: Indiana UP, 1997. 42–48.
Dean, Gabrielle. "The "Phallacies" of Dyke Comic Strips." *The Gay '90's: Disciplinary and Interdisciplinary Formations in Queer Studies*. Ed. Thomas Foster, Carol Siegel, and Ellen Berry. New York: New York UP, 1997. 199–223.
Degrassi: The Next Generation. Created by Yan Moore. CTV, 2001–.

Desert Hearts. Dir. Donna Deitch. Perf. Helen Shaver, Patricia Charbonneau, and Audra Lindley. Desert Hearts Productions, 1985.
DeVille, Willy/Mink. "Savoir Faire." *Savoir Faire.* Capitol, 1981.
Devor, Holly. *FTM: Female-to-Male Transsexuals in Society.* Bloomington: Indiana UP, 1997.
———. *Gender Blending: Confronting the Limits of Duality.* Bloomington: Indiana UP, 1989.
DiMassa, Diane. *The Complete Hothead Paisan: Homicidal Lesbian Terrorist.* San Francisco: Cleis Press, 1999.
"Director's Interview." DVD extras. *The Crash Pad.* Dir. Shine Louise Houston. Perf. Dylan Ryan, Roxie Ryder, Dusty Ryder, Jo, Jiz Lee, and Shawn. Pink and White Productions, 2005.
Dirty Pictures. Dir. Frank Pierson. Perf. James Woods, Craig T. Nelson, and Diana Scarwid. MGM Television Entertainment, 2000.
Diva [periodical]. London: Millivres Prowler Ltd., 1994–.
Doan, Laura. "Passing Fashions: Reading Female Masculinities in the 1920s." *Feminist Studies* 24 (1998): 663–700.
Dolan, Jill. "Desire Cloaked in a Trenchcoat." *The Drama Review: A Journal of Performance Studies* 31.1 (1989): 59–67.
Douglas, Erin. "That's Not Me: Queer Performance's 'Troubling' of the Desire for Authenticity in *The L-Word.*" Conference paper. Midwest Modern Language Association annual conference, St. Louis, November 2004. 4 Jul. 2005.
———."Femme Fem(me)ininities: A Performative Queering." MA Thesis. Miami U, 2004. 48–66.
———. "Pink Heels, Dildos and Erotic Play: The (Re)Making of Fem(me)ininity in Showtime's *The L Word.*" *Televising Queer Women: A Reader.* Ed. Rebecca Beirne. New York: Palgrave Macmillan, 2007. 195–209.
Drag Kings on Tour. Dir. Sonia Slutsky. Alliance Atlantis, 2004.
Duggan, Lisa. "The New Homonormativity: The Sexual Politics of Neoliberalism." *Materializing Democracy: Toward a Revitalized Cultural Politics.* Ed. Russ Castronovo and Dana D Nelson. Durham, NC: Duke UP, 2002. 175–94.
The D Word. Created by Dasha Snyder. Dyke TV, 2005. <http://www.thedword.com>.
Dyke TV. Created by Linda Chapman, Ana Simo, and Mary Patierno. Free Speech TV, 1993–.
———. Home page. 15 Aug. 2005. <http://www.dyketv.org/>.
Dyke TV. Created by Caroline Spry. Channel 4, 1995–1996.
Ebb, Fred, John Kander, Joel Grey, and Liza Minnelli. "Money Money." *Cabaret: Original Soundtrack Recording (1972 Film).* Hip-O Records, 1996.
Ellen. Created by David S. Rosenthal. ABC, 1994–1998.
Ellis, Warren. *The Authority.* 2000–.
———. *The Authority Volume One: Relentless.* New York: Wildstorm (DC Comics), 2000.
Epstein, Rachel. "Lesbian Parenting: Grounding Out Theory." *Canadian Woman Studies* 16.2 (1996): 60–64.

Emig, Rainer. "Queering the Straights, Straightening Queers: Commodified Sexualities and Hegemonic Masculinity." *Subverting Masculinity: Hegemonic and Alternative Versions of Masculinity in Contemporary Culture*. Ed. Russell West and Frank Lay. Amsterdam: Rodopi, 2000. 207–26.

Engel, Maureen. "Arousing Possibilities: The Cultural Work of Lesbian Pornography." Diss. U of Alberta, 2003.

E.R. Created by Michael Crichton. NBC, 1994–.

Erotic in Nature. Dir. Cristen Lee Rothermund. Perf. Kit Marseilles and Chris Cassidy. Tigress Productions, 1985.

Eskridge, William. *The Case for Same-Sex Marriage: From Sexual Liberty to Civilized Commitment*. New York: Free Press, 1996.

Exes and Ohs. Created by Michelle Paradise. Logo, 2007-.

Experiment: Gay and Straight. Hosts Darlene Hill and Mark Saxenmeyer. Fox Chicago, 2002.

Farrar, Stacy. "The L Word." *Sydney Star Observer* 1 Jul. 2004. 24 Jul. 2004. <http://www.ssonet.com.au/display.asp?ArticleID=3313>.

Feinberg, Leslie. *Stone Butch Blues*. Ithaca: Firebrand Books, 1993.

———. *Trans Liberation: Beyond Pink or Blue*. Boston: Beacon Press, 1999.

Fingersmith. Dir. Aisling Walsh. Adapted by Peter Ransley. BBC, 2005.

Fox, Katrina. "Blush: Making Our Own Erotica." *Lesbians on the Loose* Jul. 2004: 12–13.

Fox, Tricia. "Lesbian Wedlock." *Lesbians on the Loose* Aug. 2005: 10–11.

Franson, Leanne. *Liliane, Bi-Dyke*. 1992–.

———. "Anti-Porn Fanatic—Part 1." *Liliane* mini-comic #22. Self-published Nov. 1993. Rpt. 2004.

———. "Anti-Porn Fanatic—Part 2." *Liliane* mini-comic #26. Self-published Oct. 1994. Rpt. 2004.

———. *Assume Nothing: Evolution of a bi-dyke*. Hove, UK: Slab-O-Concrete Productions. 1997.

———. *Teaching Through Trauma*. Hove, UK: Slab-O-Concrete Productions. 1999.

Fraser, M. L. "Zine and Heard: Fringe Feminism and the Zines of the Third Wave." *Feminist Collections* 23.4 (2002): 6–10.

Freeman, Elizabeth. "Packing History, Count(er)ing Generations." *New Literary History* 31.4 (2000): 727–44.

Friends. Created by David Crane and Marta Kauffman. NBC, 1994–2004

Garber, Linda. *Identity Poetics: Race, Class, and the Lesbian-Feminist Roots of Queer Theory*. New York: Columbia UP, 2001.

Gay Comix [comics periodical]. Milwaukee: Kitchen Sink, 1980–1998. Known as *Gay Comics* after 1991.

Gay League. "Interview: Joan Hilty." *Gay League* web site. c. 2001. 24 Aug. 2005. <http://www.gayleague.com/forums/display.php?id=110>.

———. "LGBT Comics Timeline." *Gay League* web site. 19 Apr. 2005. <http://www.gayleague.com/gay/timeline/index.php>.

Gay Media Coalition. Letter/circular. New York, 17 Nov. 1975. Located in Lesbian Herstory Archives, Brooklyn, NY, Sept. 2004.

Gay Media Project. News Release. "Gays to Probe Anti-homosexual Bias on New WPVI Public Affairs Program." Contacts given: Loretta DeLoggio and John Wiles. Date unknown, c. 1970s. Located in Lesbian Herstory Archives, Brooklyn, NY, Sept. 2004.

Gelder, Ken. *Reading the Vampire*. 1994. London and New York: Routledge, 2001.

"'Get This Filth Off Our Screens.'" [Article on Queer Characters on Television]." *BBC Tipping the Velvet* Web site. Dec. 2003. 14 Aug. 2005. <http://www.bbc.co.uk/drama/tipping/article_1.shtml>.

Gever, Martha. *Entertaining Lesbians: Celebrity, Sexuality, and Self-Invention*. New York: Routledge, 2003.

Girlfriends Magazine [periodical]. San Francisco: H.A.F Publishing, 1994–.

Glitz, Michael. "Beyond *Queer as Folk*: From News to Sitcoms, Out Gays and Lesbians Are Making Their Mark All over British TV." *The Advocate* 5 Feb. 2002: 44–45.

Glock, Alison. "She Likes to Watch" [Interview with Ilene Chaiken]. *New York Times* 6 Feb. 2005: 26, 38.

Goldfish Memory. Dir. Elizabeth Gill. Perf. Sean Campion and Flora Montgomery. Goldfish Films, 2003.

Goldstein, Richard. *Homocons: The Rise of the Gay Right*. London and New York: Verso, 2003. Rpt. of *The Attack Queers: Liberal Society and the Gay Right*. 2002.

Gregory, Roberta. *Naughty Bits*. 1991–2004.

———. *Bitchy Butch: World's Angriest Dyke!* Seattle: Fantagraphics, 1999.

Gross, Larry. *Up from Invisibility: Lesbians, Gay Men, and the Media in America*. New York and Chichester: Columbia UP, 2001.

Grosz, Elizabeth. "Motherhood." *Sexual Subversions: Three French Feminists*. St. Leonards: Allen and Unwin, 1989. 78–85.

Gurlesque. "About Us." *Gurlesque Burlesque Lezzo Strip Club* Web site. 19 Nov. 2004. 21 Nov. 2005 <http://www.gurlesque.com/about.html>.

Halberstam, Judith "Jack." *Female Masculinity*. Durham, NC: Duke UP, 1998.

———. "The I Word: 'I' for Invisible, as in Real-World Lesbians on TV." *Girlfriends* Feb. 2004: 18.

Halberstam, Judith, and C. Jacob Hale. "Butch/FTM Border Wars: A Note on Collaboration." *GLQ* 4.2 (1998): 283–85.

Halberstam, Judith, "Jack," and Del LaGrace Volcano. *The Drag King Book*. London: Serpent's Tail, 1999.

Haley, Kathy. "In from the Cold." *Multichannel News* 26.26 (2005): 25.

Hall, Radclyffe. *The Well of Loneliness*. 1928. London: Virago, 2004.

Handy, Bruce. "He Called Me Ellen DeGenerate?" *Time* 14 Apr. 1997: 86.

———. "Roll Over, Ward Cleaver." *Time* 14 Apr. 1997: 78–85.

Hantzis, Darlene, and Valerie Lehr. "Whose Desire? Lesbian (Non)Sexuality and Television's Perpetuation of Hetereo/Sexism." *Queer Words, Queer Images: Communication and the Construction of Homosexuality*. Ed. Jeffrey Ringer. New York: New York UP, 1994. 107–21.

Hardie, Melissa Jane. "Beard." *Rhetorical Bodies*. Ed. Jack Selzer and Sharon Crowley. Madison: University of Wisconsin Press, 1999. 275–96.

Hard Love and How to Fuck in High Heels. Dir. Jackie Strano and Shar Rednour. Perf. Jamie Ben-Azay, Nicole Katler, and Edrie Schade. S.I.R Video, 2000.
Harrington, C. Lee. "Lesbian(s) on Daytime Television: The Bianca Narrative on All My Children." *Feminist Media Studies* 3.2 (2003): 207–28.
Hastie, Nicki. "It All Comes Out in the Wash: Lesbians in Soaps." *Trouble & Strife* 29/30 (1994/1995): 33–38.
Hay Fever. Dir. Christen Lee Rothermund. Perf. Ethyl Supreme, Jezebel Tartini, and Nina Hartley. Tigress Productions, 1989.
Healey, Emma. *Lesbian Sex Wars*. London: Virago, 1996.
Heartbeat. Created by Harry Winer. ABC, 1988–1989.
Heath, Desmond. "An Investigation into the Origins of a Copious Vaginal Discharge During Intercourse: 'Enough to Wet the Bed'—That 'Is Not Urine.'" *The Journal of Sex Research* 20.2 (1984): 194–215.
Heffernan, Virginia. "It's February; Pucker Up, TV Actresses." *The New York Times* 10 Feb. 2005. 1 Feb. 2006. <http://www.nytimes.com/2005/02/10/arts/television/10heff.html?ex=1265778000&en=81966f6411ba81ad&ei=5090&partner=rssuserland>.
Helburg, Michele. "Comics Offer Fun, Fully-Developed Lesbian, Bi Characters." *Afterellen.com*. 4 Apr. 2005. 4 Apr. 2005. <http://www.afterellen.com/Print/2005/4/comics.html>.
Heller, Dana. "Hothead Paisan: Clearing a Space for a Lesbian Feminist Folklore." *New York Folklore* 19.1–2 (1993): 27–44.
———. "Purposes: An Introduction." *Cross-Purposes: Lesbians, Feminists, and the Limits of Alliance*. Ed. Dana Heller. Bloomington and Indianapolis: Indiana UP, 1997. 1–16.
———. "States of Emergency: The Labours of Lesbian Desire in *ER*." *Genders* 39 (2004). 7 Sep. 2005. <http://www.genders.org/g39/g39_heller.html>.
Hemmings, Clare. "Out of Sight, Out of Mind? Theorizing Femme Narrative." *Sexualities* 2.4 (1999): 451-464.
Henderson, Lisa. "Lesbian Pornography: Cultural Transgression and Sexual Demystification." *New Lesbian Criticism: Literary and Cultural Readings*. Ed. Sally Munt. New York: Columbia UP, 1992. 173–91.
Herman, Didi. "'Bad Girls Changed My Life': Homonormativity in a Women's Prison Drama." *Critical Studies in Media Communication* 20.2 (2003): 141–59.
Hidalgo, Melissa. "'Going Native on Wonder Woman's Island': The Exoticization of Lesbian Sexuality in *Sex and the City*." *Televising Queer Women: A Reader*. Ed. Rebecca Beirne. New York: Palgrave Macmillan, 2007. 121–33.
The Hidden History of Homosexual Australia. Dir. Con Anemogiannis. Narrated by Simon Burke. Fortian Productions, 2004.
Hilty, Joan. *Bitter Girl*. Q Syndicate, 2003–. 1 Sep. 2005. <http://www.gmax.co.za/play/cartoons/cartoon2.html>.
———. "Immola and the Luna Legion." *Oh . . . Her Comic Quarterly*. 1991–c. 1996.
Hinds, Hilary. "*Oranges Are Not the Only Fruit*: Reaching Audiences Other Lesbian Texts Cannot Reach." *Immortal Invisible: Lesbians and the Moving Image*. Ed. Tamsin Wilton. London: Routledge, 1995. 52–69.

Holiday, Billie. "Love for Sale." 1945. *Love for Sale*. Prestige/Verve, 1993.
Hollibaugh, Amber, and Cherríe Moraga. "What We're Rollin' around in Bed With: Sexual Silences in Feminism." *Desire: The Politics of Sexuality*. Ed. Ann Snitow, Christine Stansell, and Sharon Thompson. London: Virago Press, 1984. 404–14. Rpt. of *Powers of Desire: The Politics of Sexuality*. 1983.
"*Hothouse* Talks to *Dykes to Watch Out For* Creator Alison Bechdel." *Hothouse: Blogging for Lesbians* 17 Oct. 2005. 4 Feb. 2006. <http://hthse.com/wordpress/2005/10/17/hothouse-talks-to-dykes-to-watch-out-for-creator-alison-bechdel>.
Howes, Keith. *Broadcasting It: An Encyclopaedia of Homosexuality on Film, Radio, and TV in the UK 1923–1993*. London and New York: Cassell, 1993.
———. "Gays of Our Lives: 30 Years of Gay Australian TV." *OutRage* 177 (Feb. 1998): 38–49.
How to Female Ejaculate. Dir. Nan Kinney. Perf. Fanny Fatale, Shannon Bell, and Carol Queen. Fatale Video, 1992.
Hughes, Natalya. "Gurlesque: Options Nightclub, Brisbane 7 June, 2003." *Local Art* 6 (2003): 1–2.
If These Walls Could Talk 2 [telemovie]. Dir. Jane Anderson, Martha Coolidge, and Anne Heche. Perf. Vanessa Redgrave, Michelle Williams, and Ellen DeGeneres. HBO, 2000.
ILGA-Europe. *After Amsterdam: Sexual Orientation and the European Union [online book]*. Brussels: ILGA-Europe, 1999. 15 Aug. 2005. <http://www.ilga-europe.org/>.
———. "Legal Recognition of Same-Sex Partnerships in Europe." *ILGA-Europe* Web site. 2005. 15 Aug. 2005. <http://www.ilga-europe.org/>.
Intents, Sex & Supernova, Glita. Personal Interview. 17 Jan. 2005. [Digital recording thereof held by author].
Inness, Sherrie. *The Lesbian Menace: Ideology, Identity, and the Representation of Lesbian Life*. Amherst: University of Massachusetts Press, 1997.
Jaarsma, Ada and Pederson, Tara. "An Existential Look at *The L-Word*." Conference paper. Midwest Modern Language Association annual conference, St. Louis, November 2004. 4 Jul. 2005.
Jeffreys, Sheila. "How Orgasm Politics Has Hijacked the Women's Movement." *On the Issues: The Progressive Woman's Quarterly* Spring 1996. 16 Jun. 2005. <http://www.ontheissuesmagazine.com/s96orgasm.html>.
Jensen, Steffen. "Denmark." *Equality for Lesbians and Gay Men: A Relevant Issue in the Civil and Social Dialogue*. Ed. Nico J. Bergo, Kurt Krickler, Jackie Lewis, and Maren Wuch. Brussels: ILGA-Europe, 1998. 38–41. 15 Aug. 2005. <http://www.steff.suite.dk/report.pdf>.
Jetter, Alexis. "Goodbye to the Last Taboo." *Vogue* July 1993: 86–88, 92.
Johns, Merryn. "Wedlock." *Lesbians on the Loose* July 2004: 8.
Johnson, Margaret E. "Boldly Queer: Gender Hybridity on *Queer as Folk*." *Quarterly Review of Film and Video* 21 (2004): 293–301.
Johnson, Peter. "Salt Shakers and 'The L Word': Censorship by Subterfuge." *Libertus.net* 9 Jul. 2004. 22 Aug. 2005 <http://libertus.net/censor/odocs/pj-lword.html>.

Kaiser, Charles. "*Queer as Folk*—That's a Wrap." *Planetout Entertainment* web site. 12 Apr. 2005. 12 Apr 2005. <http://www.planetout.com/entertainment/news/?sernum=1015>.

Kamentsky, Gina. *T-Gina*. 1997–. 10 Jul. 2006. <http://www.t-gina.com/Comic/comics.html>.

Keehnen, Owen. "Well Drawn: Talking with Cartoonist Alison Bechdel." *GLBTQ: An Encyclopaedia of Gay, Lesbian, Bisexual, Transgender, and Queer Culture* Web site. 1994. 24 Aug. 2005. <http://www.glbtq.com/sfeatures/interviewabechdel.html>.

Keller, James R. *Queer (Un)Friendly Film and Television*. Jefferson, NC: McFarland, 2002.

Keller, James R., and Leslie Stratyner. *The New Queer Aesthetic on Television: Essays on Recent Programming*. Jefferson, NC: McFarland, 2006.

Keller, Yvonne. "Pulp Politics: Strategies of Vision in Pro-Lesbian Pulp Novels, 1955–1965." *The Queer Sixties*. Ed. Patricia Juliana Smith. New York: Routledge, 1999. 1–25.

"King for a Night" [Interview with Sexy Galexy]. *Lesbians on the Loose* Aug. 2005. 18.

Kinnard, Rupert. *Cathartic Comics* (The Brown Bomber and Diva Touché Flambé). 1977–1987.

———. *B.B. and the Diva*. Boston: Alyson Publications, 1992.

BurLEZk Live! Dir. Nan Kinney and Debbie Sundahl. Perf. Ramona, Pepper and Sandra. Blush Entertainment Group/Fatale Video, 1986.

Kissing Jessica Stein. Dir. Charles Herman-Wurmfeld. Perf. Jennifer Westfeldt, Heather Juergensen, and Scott Cohen. Fox Searchlight Pictures, 2001.

Kregloe, Karman. "Iyari Limon on Buffy, Bisexuality, and Adventures in Cooking." *Afterellen.com*. 10 Apr. 2006. 18 Apr. 2006. <http://www.afterellen.com/People/2006/4/iyari.html>.

Kristen. "Rundown" Firsthand account of "No Limits: A Look at *Queer as Folk* and *The L Word*," a seminar at The Museum of Television & Radio, New York, 24 May 2004. *The L Word Online* fan-site. 13 Jun. 2004. 24 Jul. 2004. <http://www.thelwordonline.com/MTR_rundown.shtml>.

L.A. Law. Created by Steven Bochco and Terry Louise Fisher. NBC, 1986–1994.

Law and Order. Created by Dick Wolf. NBC, 1990–.

Lee, Sadie. "Raging Bull." 1992. Private Collection. *Gallery Three—Tomboys and Crossdressers*. Artist's web site, Fig. 1. 20 Apr. 2006. <http://www.sadielee.f9.co.uk/gallery_three.htm>.

———. "La Butch en Chemise." 1992. Private Collection. *Gallery Three—Tomboys and Crossdressers*. Artist's web site, Fig. 2. 20 Apr. 2006. <http://www.sadielee.f9.co.uk/gallery_three.htm>.

———. "Venus Envy." 1994. For sale. *Gallery Three—Tomboys and Crossdressers*. Artist's web site, Fig. 8. 20 Apr. 2006. <http://www.sadielee.f9.co.uk/gallery_three.htm>.

"Lesbian Chic: The Bold, Brave New World of Gay Women" [cover]. *New York* May 1993.

Lesbians on the Loose [periodical]. Darlinghurst: Avalon Media Pty Ltd., 1989–.

Levy, Ariel. "Where the Bois Are: Why Some Young Lesbians Are Going beyond Feminist Politics, beyond Androgyny, to Explore a New Generation of Sex Roles." *New York* 12 Jan. 2004. 19 Jan. 2004. <http://www.newyorkmetro.com/nymetro/news/features/n_9709//index.html>.

Lezzie Smut [periodical]. Canada, 1993–c. 1998.

Lewis, Reina, and Katrina Rolley. "Ad(dressing) the Dyke: Lesbian Looks and Lesbians Looking." *Outlooks: Lesbian and Gay sexualities and visual cultures*. Ed. Peter Horne and Reina Lewis. London and New York: Routledge, 1996. 178–90.

Liepe-Levinson, Katherine. "Striptease: Desire, Mimetic Jeopardy, and Performing Spectators." *The Drama Review* 42.2 (1998): 9–37.

Lo, Malinda. "Gender Trouble on *The L Word*." *Afterellen.com*. 6 Apr. 2006. 15 May 2007. <http://www.afterellen.com/TV/2006/4/butches.html>.

———. "Interview with *The L Word*'s Daniela Sea." *Afterellen.com*. 3 Jan. 2006. 15 May 2007. <http://www.afterellen.com/TV/2006/1/sea.html>.

———. "It's All About the Hair: Butch Identity and Drag on *The L Word*." *Afterellen.com*. Apr. 2004. 1 May 2004. <http://www.afterellen.com/TV/thelword/butch.html>.

———. "Lesbian and Bisexual Women on *Law and Order*." *Afterellen.com*. 13 Sep. 2005. 2 Nov. 2005. <http://www.afterellen.com/TV/2005/9/lawandorder.html>.

———. "She's a Very Dirty Girl: The Transformation from Art Student to Pornographer Was a Simple One for Queer Auteur Shine Louise Houston." *Curve* Feb. 2006: 42–43.

———. "'Woman Power' on *Survivor: Vanuatu*." *Afterellen.com*. 22 Nov. 2004. 3 Feb. 2005. <http://www.afterellen.com/TV/112004/survivor.html>.

Lorde, Audre. *Sister Outsider: Essays and Speeches*. Trumansburg, NY: Crossing Press, 1984.

The Love Boat. Created by Aaron Spelling. ABC, 1977–1986.

Lumby, Catherine. "Sexist or Sexy?: Feminism's New Wowsers." *The Independent Monthly* Nov. 1993: 30–35.

The L Word. Created by Ilene Chaiken. Showtime, 2004–.

"L Word Fashion Extra." *The L Word: Season One*. Showtime Entertainment, 2004. DVD, Disc 5.

"*The L Word*: Season 2 and 3 News." *Afterellen.com*. 15 Feb. 2005. 19 Jul. 2005. <http://www.afterellen.com/TV/thelword-news.html>.

"'L Word' to fill SATC style gap?" *Fashionweekdaily.com*. 13 May 2004. 14 Nov. 2005. <http://www.fashionweekdaily.com/news/fullstory.sps?iNewsID=206807&itype=8486>.

Mack, Kelly. "Unveiling Alison Bechdel." *The College News* 3 Feb 1999. 24 Aug. 2005. <http://www.brynmawr.edu/orgs/cnews/020300/bechdel.html>.

Madam and Eve. Dir. Angie Dowling (a.k.a. Rusty Cave). Perf. Bunni Noir, Ms Jones, Milly Lane, and Sheba Ballingdon. Rusty Films, 2003.

Making Grace. Dir. Catherine Gund. Documentary. Feat. Ann Krusl and Leslie Sullivan. Aubin Pictures, 2005.

Mangels, Andy. "Out in Comics." *The Pride* 2 (2000). 17 Aug. 2005. <http://www.gayleague.com/gay/history/outcomicsbrief.php>.
Mango Kiss. Dir. Sascha Rice. Perf. Daniele Ferraro, Michelle Wolff, and Sally Kirkland. Mango Me Productions, 2004.
Marshment, Margaret, and Julia Hallam. "Framing Experience: Case Studies in the Reception of *Oranges Are Not the Only Fruit*." *Screen* 36 (1995): 1–15.
———."From String of Knots to Orange Box: Lesbianism on Prime Time." *The Good, the Bad, and the Gorgeous: Popular Culture's Romance with Lesbianism*. Ed. Diane Hamer and Belinda Budge. London: Pandora, 1994. 142–65.
Martin, Biddy. *Femininity Played Straight: The Significance of Being Lesbian*. New York and London: Routledge, 1996.
———. "Sexualities without Genders and Other Queer Utopias." *Diacritics* 24.2–3 (1994): 104–21.
Martindale, Kathleen. *Un/popular Culture: Lesbian Writing after the Sex Wars*. Albany: State University of New York Press, 1997.
McCroy, Winnie. "'L' is for Invisible." *New York Blade* 31 Oct. 2003. 25 Jul. 2005. <http://www.thelwordonline.com/L_is_for.shtml>.
McMillan, Jackie. "Live at the Hellfire Club." *Slit* 8 (2005): 52–53.
Meatmen [comics periodical]. San Francisco: Leyland Publications, 1986–.
Merck, Mandy. "Desert Hearts." *Queer Looks: Perspectives on Lesbian and Gay Film and Video*. New York: Routledge, 1993. 377–82.
Meredith. Radio interview, Sydney Indymedia Centre. 2 Mar. 2001. 20 Feb. 2005. <http://sydney.indymedia.org/radio.php3>.
Meredith and Domino. "Wicked Women Revisited." *Slit* 2 (2002): 8–9.
———. "Gurlesque Lesbian Strip Club: Talking with Sex and Glita." *Slit* 2 (2002): 14–18.
Metropolis S.C.U. [comic book series]. New York: DC Comics, 1995–1996.
Moen, Erika. "Miscellaneous comics." *Projectkooky.com*. 2000–. 10 Jul. 2006. <http://www.projectkooky.com/erika/comics/>.
Monteiro, Kate, and Sharon Bowers. "Too Queer for *Queer as Folk*: Identity, Patriarchy, and the Lesbian Subject." Paper presented at the Southwest Popular Culture and American Culture Associations' Annual Conference. Oct. 2002. Rpt. online. 5 Sept. 2003. <http://sharonbowers.com/GasNSip/QAFPaper.htm>.
Moore, Booth. "Dressed-up Diversity: On 'The L Word,' Lesbians Go Chic, Playing against Stereotypes but Raising Issues of Identity and Acceptance." *The LA Times*. 8 Feb. 2004. 15 Mar 2004. <http://www.after0ellen.com/About/Press/latimes-feb2004.html>.
Moore, Candace, and Kristen Schilt. "Is She Man Enough?: Female Masculinities on *The L Word*." *Reading The L Word: Outing Contemporary Television*. Ed. Kim Akass and Janet McCabe. London: I. B. Tauris, 2006. 159–71.
Moorman, Jennifer. "'Shades of Grey': Articulations of Bisexuality in *The L Word*." *Televising Queer Women: A Reader*. Ed. Rebecca Beirne. New York: Palgrave Macmillan, 2007. 163–76.
Moritz, Marguerite. "Old Strategies for New Texts: How American Television Is Creating and Treating Lesbian Characters." *Queer Words, Queer Images: Communication*

and the Construction of Homosexuality. Ed. Jeffrey Ringer. New York: New York UP, 1994. 122–42.

Mulvey, Laura. "Afterthoughts on 'Visual Pleasure and Narrative Cinema' inspired by King Vidor's *Duel in the Sun* (1946)" 1981. Rpt. in *Visual and Other Pleasures*. Laura Mulvey. Houndsmills, Basingstoke, Hampshire: The Macmillan Press, 1989. 29–38.

———. "Visual Pleasure and Narrative Cinema." *Screen* 16.3 (1975): 6–18.

Munt, Sally, ed. *Butch/Femme: Inside Lesbian Gender*. London: Cassell, 1998.

———. "Shame/Pride Dichotomies in *Queer as Folk*." *Textual Practice* 14.3 (2000): 531–46.

My Best Friend's Wedding. Dir. P. J. Hogan. Perf. Julia Roberts, Dermot Mulroney, and Rupert Everett. Tristar Pictures, 1997.

Namaste, Viviane K. "Interview with Mirha Soleil-Ross." *Sex Change, Social Change: Reflections on Identity, Institutions, and Imperialism*. Toronto: Women's Press, 2005. 86–102.

———. *Invisible Lives: The Erasure of Transexual and Trangendered People*. Chicago and London: University of Chicago Press, 2000.

Nataf, Zachary I. *Lesbians Talk Transgender*. London: Scarlett Press, 1996.

Neighbours. Created by Reg Watson. Channel 10, 1985–.

Nestle, Joan. "The Fem Question." *Pleasure and Danger: Exploring Female Sexuality*. Ed. Carole S. Vance. Boston, London, Melbourne, and Henley: Routledge and Kegan Paul, 1984. 232–41.

———. Introduction. *A Restricted Country*. 1987. San Francisco: Cleis Press, 2003.

Nestle, Joan, Riki Wilchins, and Clare Howell. *Genderqueer: Voices from Beyond the Sexual Binary*. Los Angeles: Alyson Publications, 2002.

Newman, Felice. "Don't Miss *The Crash Pad*." *San Francisco Bay Times* 2 Mar. 2006. 8 Apr. 2006. <http://www.sfbaytimes.com/?sec=article&article_id=4706>.

Newman, Lesléa. "I Enjoy Being a Girl." *The Femme Mystique*. Ed. Lesléa Newman. Boston: Alyson Publications, 1995. 11–13.

NGLTF (National Gay and Lesbian Taskforce). "Lesbian, Gay, Bisexual and Transgender (LGBT) Parents and their Children." *Thetaskforce.org* 18 Aug. 2004. 23 Aug. 2005. <http://www.thetaskforce.org/downloads/LGBTParentsChildren.pdf>.

Nguyen, Kenneth. "Soapie Kiss Sparks Outcry." *The Age* 27 Sep. 2004: 5.

Noah's Arc. Created by Patrik-Ian Polk. Logo, 2004–.

Noble, Bobby. "Queer as Box: Boi Spectators and Boy Culture on Showtime's *Queer as Folk*." *Third Wave Feminism and Television: Jane Puts It in a Box*. Ed. Merri Lisa Johnson. London: I. B. Tauris, 2007. 147–65.

"Not Your Mother's Lesbians" [cover]. *New York* Jan. 2004.

Nyberg, Amy Kiste. *Seal of Approval: The History of the Comics Code*. Jackson: UP of Mississippi, 1998.

Nyong'o, Tavia. "Queer TV: A Comment." *GLQ: A Journal of Lesbian and Gay Studies* 11.1 (2005): 103–5.

The O.C. Created by Josh Schwartz. Fox, 2003–.

On Our Backs: Entertainment for the Adventurous Lesbian. Fatale Media, 1984–1995; H.A.F. Publishing, 1998–.

Opie, Catherine. "Self-Portrait." 1993. Regen Projects, Los Angeles.
The Oprah Winfrey Show. Created by Oprah Winfrey. WLZ-TV (syndicated nationally by King World), 1985–.
Oranges Are Not the Only Fruit. Dir. Beeban Kidron. Adapted by Jeanette Winterson. BBC, 1990.
Other Mothers [telemovie]. Dir. Lee Shallat Chemel. Perf. Joanna Cassidy and Meredith Baxter. CBS, 1993.
Out. Programmed by Peter Wasson. SBS, 1993–1995.
Out of the Closet TV. *Out of the Closet TV: The World's First Broadband GLBT Broadcasting Network Since 2000.* Home page. 18 Apr. 2006. <http://www.outofthecloset.tv>.
Palmer, Joe. "Renee Montoya." *Gay League* Web site. 5 Dec. 2007. <http://www.gayleague.com/gay/characters/display.php?id=212>
Parks, Joy. "Straying From the Gender Pack: An Interview with Ivan E. Coyote." *Herizons* Summer 2003. 10 Aug. 2004. <http://www.herizons.ca/magazine/backissues/sum03/profile.html>.
Peeren, Esther. "Queering the Straight World: The Politics of Resignificaton in *Queer as Folk.*" *The New Queer Aesthetic on Television: Essays on Recent Programming.* Ed. James R. Keller and Leslie Stratyner. Jefferson, NC: McFarland, 2006. 59–74.
Phelan, Shane. "Public Discourse and the Closeting of Butch Lesbians." *Butch/Femme: Inside Lesbian Gender.* Ed. Sally Munt. London and Washington: Cassell, 1998. 191–99.
Prisoner: Cell Block H. Created by Reg Watson. Channel 10, 1979–1986.
Pottie, Lisa. "Hierarchies of Otherness: The Politics of Lesbian Styles in the 1990s, or What to Wear?" *Canadian Woman Studies* 16.2 (1996): 49–52. Expanded Academic ASAP. Gale Group/InfoTrac. U of Sydney Libraries. 2 Nov. 2005. <http://infotrac.galegroup.com/itweb/>.
Queer as Folk. Created by Russell T. Davies. Channel 4, 1999–2000.
Queer as Folk. Created by Ron Cowen, Daniel Lipman, and Tony Jonas. Showtime, 2000–2005.
Queer Eye for the Straight Girl. Created by David Collins. Bravo, 2005–.
Queer Eye for the Straight Guy. Created by David Collins. Bravo, 2003–.
Queerscreen. "2004 Mardi Gras Film Festival Guide." Liftout from *The Sydney Star Observer* 15 Jan. 2004. QS-01–QS-20.
Quim [periodical]. United Kingdom, 1989–1995.
Raikas, Tuula. "Humour in Alison Bechdel's Comics." Pro Gradu Thesis (Masters, Faculty of Humanities). U of Jyväskylä, 1997.
Rasmussen, Mary-Lou and Kenway, Jane. "Queering the Youthful Cyberflaneur." *Journal of Gay and Lesbian Issues in Education* 2.1 (2004): 47-63.
Reeder, Constance. "The Skinny on the L Word." *Off Our Backs* 34.1–2 (2004): 51–52.
Renshaw, Sal and Robinson, Laura M. "Where *The L-Word* Meets the F Word: Random Acts and the Limits of Representation." Conference paper. Midwest Modern Language Association annual conference, St. Louis, November 2004. 4 Jul. 2005.
Richards, Lisa. "The 9 Lives of Kitten" [Interview]. *Diva* Aug. 2004: 29–30.

Richardson, Diane. "'Misguided, Dangerous and Wrong': On the Maligning of Radical Feminism." *Radically Speaking: Feminism Reclaimed.* Ed. Diane Bell and Renate Klein. North Melbourne: Spifex, 1996. 143–54.
Robbins, Trina. *From Girls to Grrrlz: A History of Women's Comics from Teens to Zines.* San Francisco: Chronicle Books, 1999.
Robinson, Paul. *Queer Wars: The New Gay Right and Its Critics.* Chicago and London: University of Chicago Press, 2005.
Roof, Judith. *Come as You Are: Sexuality and Narrative.* New York and Chichester: Columbia UP, 1996.
Roseanne. Created by Matt Williams. ABC, 1988–1997.
Rowlson, Alex. "Gets Real." *Fab: The Gay Scene Magazine* 2004. 12 Mar 2006. <http://www.fabmagazine.com/features/insideout2004/>.
Russo, Vito. *The Celluloid Closet: Homosexuality in the Movies.* New York: Harper and Row, 1987.
Safe Is Desire. Dir. Debi Sundahl. Perf. Christine Lareina, Darby Michael, Elizabeth Marshall, and Karen Everett. Blush/Fatale, 1993.
San Martin, Nancy. "Queer TV: Framing Sexualities on US Television." Diss. U of California, Santa Cruz, 2002.
Scalettar, Liana. "Resistance, Representation, and the Subject of Violence: Reading *Hothead Paisan.*" *Queer Frontiers: Millenial Geographies, Genders, and Generations.* Ed. Joseph A. Boone, Martin Dupais, Martin Meeker, Karen Quimby, Cindy Sarver, Debra Silverman, and Rosemary Weatherston. Madison and London: The U of Wisconsin P, 2000. 261–77.
Schilt, Kristen. "A Little Too Ironic: The Appropriation and Packaging of Riot Grrrl Politics by Mainstream Female Musicians." *Popular Music and Society* 26.1 (2003): 5–16.
Sedgwick, Eve Kosofsky. "*The L Word*: Novelty in Normalcy." *The Chronicle of Higher Education* 16 Jan. 2004: B10–B11.
Seidman, Steven. *Beyond the Closet: The Transformation of Gay and Lesbian Life.* New York: Routledge, 2002.
Selke, Lori. "Dykes in Comics." *Curve Magazine* 15.4 (June 2005): 71.
Serving in Silence: The Margarethe Cammermeyer Story [telemovie]. Dir. Jeff Bleckner. Perf. Glenn Close and Judy Davis. HBO, 1995.
Sevely, J. Lowndes, and J. W. Bennett. "Concerning Female Ejaculation and the Female Prostate." *The Journal of Sex Research* 14.1 (1978): 1–20.
"Sexy Screensaver Lands Police Officer in Hot Water." *ChannelCincinnati.com.* 16 Mar. 2005. 20 Jul. 2005. <http://www.channelcincinnati.com/news/4291756/detail.html>
Shaw, Justine. *Nowhere Girl.* 19 Apr. 2005. <http://www.nowheregirl.com/>.
Shister, Gail. "'L Word' Is Good Word for Series' Local Star." *Philadelphia Inquirer* 3 Feb. 2004. 15 Mar 2004. <http://www.philly.com/mld/philly/entertainment/television/7859765.htm>.
Showtime Networks Inc. "More Love! More Lust! More Longing!" [Press Release]. *Thelwordonline.com.* 2 Feb. 2006. 20 Apr. 2006. <http://www.thelwordonline.com/season4_renewal.html>.

Simone, Nina. "I Want a Little Sugar in My Bowl." *Nina Simone Sings the Blues*. RCA Victor, 1967.

Slit [periodical]. Broadway: Stealth and Domino, 2002–.

Snow, Judith E. *How It Feels to Have a Gay or Lesbian Parent: A Book by Kids for Kids of All Ages*. Binghamton, NY: Harrington Park Press, 2004.

Some Like It Hot. Dir. Billy Wilder. Perf. Jack Lemmon, Tony Curtis, and Marilyn Monroe. Metro Goldwyn Meyer, 1959.

Sorority Boys. Dir. Wallace Wolodarsky. Perf. Barry Watson, Michael Rosenbaum, and Harland Williams. Touchstone Pictures, 2002.

South of Nowhere. Created by Tom Lynch. The N, 2005–.

Stacey. *Welcome to Sunny Camp Trans* [Comic/Zine]. Boston: Butch Dyke Boy Press and the Boston Lesbian Avengers, 2000. <http://www.butchdykeboy.com/zines/wtsct/ppframe.htm>.

Stein, Arlene. "All Dressed Up, but No Place to Go? Style Wars and the New Lesbianism." *The Persistent Desire: A Femme-Butch Reader*. Ed. Joan Nestle. Boston: Alyson Publications, 1992. 431–40.

———. *Sex and Sensibility: Stories of a Lesbian Generation*. Berkeley, Los Angeles, and London: University of California Press, 1997.

Steinem, Gloria. "Erotica vs. Pornography." 1977/78. *Outrageous Acts and Everyday Rebellions*. 1983. New York: Henry Holt and Company, 1995. 238–51.

Stevens, Genia V. "*SistersTalk* Interviews *Dykes to Watch Out For* Creator Alison Bechdel." *SistersTalk* Blog/Web site. 4 Apr. 2005. 14 Jun. 2005. <http://sisterstalk.com/alisonbechdel.html>. Comments at "0" <http://www.tblog.com/templates//index.php?bid=sisterstalk&static=433516>.

Stockwell, Anne. "Soul of *The L Word*." *The Advocate* 17 Jan. 2006. 20 Apr. 2006. <http://www.advocate.com/print_article_ektid24451.asp>.

Stormé: Lady of the Jewel Box. Dir. Michelle Parkerson. Feat. Stormé DeLaverie. Eye of the Storm Productions, 1987.

Sudell, Denise. "13 Questions: Jennifer Camper." *Sequential Tart* Aug. 2005. 17 Aug. 2005. <http://www.sequentialtart.com/13quest_0805.shtml>.

Sugar High Glitter City. Dir. Shar Rednour and Jackie Strano. Perf. Simone de la Getto, Jackie Strano, Josephine X, and Shar Rednour. S.I.R. Video, 2001.

Sullivan, Andrew. "The Marriage Moment." *The Advocate* 750/751 (20 Jan. 1998): 59.

Supernova, Glita. Telephone conversation with the author. 26 Mar. 2006.

Surkan, Kim. "Drag Kings in the New Wave: Gender Performance and Participation." *The Drag King Anthology*. Ed. Donna Troka, Kathleen LeBesco, and Jean Noble. Binghamton, NY: The Haworth Press, 2002. 161–88.

Survivor. Created by Charlie Parsons. CBS, 2000–.

Taste This (Anna Camilleri, Ivan E. Coyote, Zoë Eakle, and Lyndell Montgomery). *Boys Like Her: Transfictions*. Vancouver: Press Gang Publishers, 1998.

Taylor, C. "Candace and Lesbians of Color on *The L Word*." *Afterellen.com*. Apr. 2004. 18 Aug. 2004. <http://www.afterellen.com/TV/thelword/candace.html>.

Tenko. Created by Lavinia Warner. BBC, 1981–1984.

Thalheimer, Anne. "Terrorists, Bitches, and Dykes: Gender, Violence, and Heteroideology in Late 20th-Century Lesbian Comix." Diss. U of Delaware, 2002.
Tick Tock. Dir. Angie Dowling. Perf. Diamond, El Capitino, and Crystal. Rusty Films, 2001.
"Tigress Presents 'Hay Fever': A Lesbian Erotic Cowgirl Comedy." Advertising pamphlet, 1989. Located in Lesbian Herstory Archives, Brooklyn, USA, Sept. 2004.
Tipping the Velvet. Dir. Geoffrey Sax. Adapted by Andrew Davies. BBC, 2002.
Thynne, Lizzie. "Being Seen: "The Lesbian" in British Television Drama." *Territories of Desire in Queer Culture: Refiguring Contemporary Boundaries.* Ed. David Alderson and Linda Anderson. Manchester and New York: Manchester UP, 2000. 202–12.
Tom of Finland (Touko Laaksonen). Cover. *Physique Pictorial* Spring 1957.
Torres, Sasha. "Television/Feminism: *Heartbeat* and Prime Time Lesbianism." *The Lesbian and Gay Studies Reader.* Ed. Henry Abelove, Michele Aina Barale, and David M Halperin. New York and London: Routledge, 1993. 176–85.
"Touch of Velvet." *Lesbians on the Loose* Aug. 2005: 28.
Trebay, Guy. "The Secret Power of Lesbian Style." *The New York Times.* 27 Jun. 2004. 29 Jun. 2004. <http://www.nytimes.com/2004/06/27/fashion/27LESB.html?pagewanted=3&ei=5062&en=a2e2eb5559cce8f4&ex=1088913600&partner=GOOGLE>.
Tree [witness to the Stonewall riots]. Personal interview. Stonewall Inn, New York City. 26 October 2004.
"Tribute to Channel Four: 1997 Frameline Award." 1997. *Frameline* Web site. 15 Aug. 2005. <http://www.frameline.org/festivals/archive/21st/events/channel4.html>.
Troka, Donna, Kathleen LeBasco, and Jean Noble. eds. *The Drag King Anthology.* Binghamton, NY: The Haworth Press, 2002.
Troka, Donna Jean (in consultation with Julie Applegate, Sile Singleton, and Shani Scott). "The History of the First International Drag King Extravaganza (IDKE 1)" *IDKEchicago.com.* c. 2004. 29 Aug. 2005. <http://idkechicago.com/oldidke/downloads/history.doc>.
Tropiano, Stephen. *The Prime Time Closet: A History of Gays and Lesbians on TV.* New York: Applause Theatre and Cinema Books, 2002.
"TV's Gay Heat Wave!" [cover]. *Vanity Fair.* Dec. 2003.
The Undergrad [short film]. Dir. Michele Mahony. Perf. the Chicago Kings and Diane Torr. Moe Films, 2003.
Vaid, Urvashi. *Virtual Equality: The Mainstreaming of Gay and Lesbian Liberation.* New York: Anchor, 1995.
VelvetPark [periodical]. Brooklyn: Parasmani Publishing LLC, 2002–.
Venus Boyz. Dir. Gabriel Baur. Onix Films, 2002.
Vivekananda, Kitty. "Integrating Models for Understanding Self-injury." *Psychotherapy in Australia* 7.1 (2000): 18–25.
Wainwright, Rufus. "Hallelujah." *The L Word [Soundtrack].* Tommy Boy, 2004.
Waldron, Petra, and Jennifer Finch. *The Adventures of a Lesbian College Schoolgirl.* New York: Amerotica/NBM, 1998.

Walker, Lisa. "How to Recognize a Lesbian: The Cultural Politics of Looking Like What You Are" *Signs* 18 (1993): 866–90.

———. *Looking Like What You Are: Sexual Style, Race, and Lesbian Identity*. New York: New York UP, 2001.

Walters, Suzanna Danuta. *All the Rage: The Story of Gay Visibility in America*. Chicago: University of Chicago Press, 2001.

———. "From Here to Queer: Radical Feminism Postmodernism and the Lesbian Menace (Or, Why Can't a Woman Be More Like a Fag?)" *Signs* 21.4 (1996): 830–69.

Warn, Sarah. "CBS's *Big Brother* Introduces Its First Lesbian Contestant." *Afterellen.com*. 11 Jul. 2005. 22 Aug. 2005. <http://www.afterellen.com/TV/2005/7/bigbrother6.html>.

———. "Lesbians Come Out of the Reality TV Closet in 2004." *Afterellen.com*. 13 Jul. 2004. 3 Feb. 2005. <http://www.afterellen.com/TV/122004/realityTV.html>.

———. "*Queer as Folk* Tackles Lesbians Who Sleep with Men—and Misses." *Afterellen.com*. Jul. 2004. 11 Jan. 2005. <http://www.afterellen.com/TV/qaf/season4-72004.html>.

———. "Review of *The L Word*." *Afterellen.com*. Jan. 2004. 20 Jan. 2004. <http://www.afterellen.com/TV/thelword/review.html>.

———. "Sex and *The L Word*." *Afterellen.com*. Apr. 2003. 12 May. 2003. <http://www.afterellen.com/TV/thelword-sex.html>.

———. "Too Much Otherness: Femininity on *The L Word*." *Afterellen.com*. Apr. 2004. 16 Apr. 2004. <http://www.afterellen.com/TV/thelword/femininity.html>.

———. "TV's Lesbian Baby Boom." *Afterellen.com*. Jan. 2003. 31 Aug. 2005. <http://www.afterellen.com/TV/lesbianbabyboom.html>.

———. "Vanity Fair the latest to Use Gay Men as Lesbian Equivalent." *Afterellen.com*. Nov. 2003. 15 Aug. 2005. <http://www.afterellen.com/Print/vanityfair.html>.

Warner, Michael. *The Trouble with Normal: Sex, Politics, and the Ethics of Queer Life*. 1999. Cambridge: Harvard UP, 2003.

Warner, Sara. "From Hothead to SCUM: Homicidal Lesbian Terrorists Storm the Scene." [Description of project] *Queer Research Group (ASTR)* Web site. 2004. 29 Aug. 2005. <http://www.lisa-raymond.com/ASTR/warner.htm>.

Warren, Roz, ed. *Dyke Strippers: Lesbian Cartoonists A to Z*. Pittsburgh and San Francisco: Cleis Press, 1995.

Watasin, Elizabeth. *Charm School* 1997–.

———. "Bunny the Good Li'l Teen Witch." *Action Girl* #13. San Jose, CA: Slave Labour Graphics, 18 Oct. 1997. 7–12.

———. *Charm School Book One: Magical Witch Girl Bunny*. San Jose: SLG Publishing, 2002.

Weaver, Susan. "Drawing the Line: Lesbian Border Wars." *Lesbians on the Loose* Nov. 2003. 12–13.

Weston, Kath. "Forever Is a Long Time: Romancing the Real in Gay Kinship Ideologies." *Long Slow Burn: Sexuality and Social Science*. New York and London: Routledge, 1998. 57–82.

Wherrett, Richard. *Mardi-Gras!: True Stories*. Ringwood, VIC: Viking, 1999.
Wicked Women [periodical]. Sydney: Wicked Women Publications, 1987–1996.
Will & Grace. Created by David Kohan and Max Mutchnick. NBC, 1998–2006.
Williams, Laura Anh and Jonet, M. Catherine. "Lesbians in Popular Culture: Not the 'L-Word' Anymore?" Conference paper. Midwest Modern Language Association annual conference, St. Louis, November 2004. 4 Jul. 2005.
Williams, Linda. *Hardcore: Power, Pleasure, and the Frenzy of the Visible*. Berkeley and Los Angeles: University of California Press, 1989.
Wilton, Tamsin. *Finger-Licking Good: The Ins and Outs of Lesbian Sex*. London and New York: Cassell, 1996.
Winslade, J. Lawton. "Teen Witches, Wiccans, and "Wanna-Blessed-Be's": Pop-culture Magic in *Buffy the Vampire Slayer*." *Slayage: The Online International Journal of Buffy Studies* 1 (2001). 10 Jul. 2003. <http://www.slayage.tv/essays/slayage1/winslade.htm>.
Wolfe, Susan J and Roripaugh, Lee Ann. "Witches and Femmes: Packaging Lesbians for Television." Conference paper. Midwest Modern Language Association annual conference, St. Louis, November 2004. 4 Jul. 2005.
Zambreno, Kate. "It's Ladies Night: The Chicago Kings Usher In a New Lesbian Scene." *Newcity Chicago* 25 Jun. 2003. 17 Feb. 2005. <http://www.newcitychicago.com/chicago/2621.html>.
Zimmerman, Bonnie. "'Confessions' of a Lesbian Feminist." *Cross-Purposes: Lesbians, Feminists, and the Limits of Alliance*. Ed. Dana Heller. Bloomington and Indianapolis: Indiana UP, 1997. 157–68.
———. "Lesbians Like This and That: Some Notes on Lesbian Criticism for the Nineties." *New Lesbian Criticism: Literary and Cultural Readings*. Ed. Sally Munt. New York and Oxford: Columbia UP, 1992. 1–15.
Zita, Jacquelyn N. "Male Lesbians and the Postmodernist Body." *Hypatia* 7.4 (1992): 106–27.

Index

1950s, 13, 103–4, 107
1960s, 13, 103, 140
1970s, 4, 8, 12, 35, 45, 53, 56, 117, 140, 196
1980s, 15, 18, 24, 47, 52, 55, 79, 95, 114, 133, 135, 140–41, 209
1990s
   1990s lesbian culture and personalities, 8, 15, 32, 47, 52, 54, 56, 75, 79, 201, 207
   1990s lesbian representation, 1, 11, 14, 16, 18, 24–25, 49, 58–59, 62, 95, 105, 113, 114, 133, 140, 196, 205
   1990s queer/lesbian theory, 4, 11, 15, 97

assimilationism, 1, 10, 18, 21–22, 38, 44, 63, 67–69, 72–74, 79, 81, 84, 93, 121, 185–87, 202–4

Bechdel, Alison, 19, 28–30, 34, 53, 54–56, 59, 167–92, 202, 209. *See also Dykes to Watch Out For*
bisexuality
   characters, 35–36, 41, 43–44, 58, 70, 85–90, 103, 106, 108, 176, 189, 207
   people, 21, 29, 31, 43, 113, 206, 209, 199–201, 206, 209
Buffy the Vampire Slayer, 17, 30, 37–38, 42, 58, 70, 199, 202
butch, 15, 31, 39, 62, 99, 104, 114, 196
   characters, 19, 53, 58, 75, 84, 91–92, 102–3, 105, 111, 113, 117–19, 121, 152, 173
   theorising butches, 11–14, 96, 98, 115–16, 118, 146, 153
butch-femme, 2, 16, 32, 38, 47–48, 53, 61, 90–92, 96, 103–4, 115, 121, 146, 160, 192, 193
Butler, Judith, 4, 13, 117, 136–37, 196, 205

camp, 2, 12, 15, 32, 68, 92, 96, 102, 114, 117, 142–43, 160, 164, 180–82
Case, Sue-Ellen, 6, 8–10, 13, 96, 138
Chaiken, Ilene, 96, 114, 119, 123, 126, 128, 207
Ciasullo, Ann, 14–16, 97–98, 103, 105, 113, 196, 205
class, 22, 91, 108, 164, 172, 193
   middle class, 9, 22, 102
   working class, 7, 39, 62, 103, 179
comics, 4, 19, 53–59, 148, 167–90, 197–98, 201, 209
coming out, 30, 37–38, 50, 79, 87–88, 100–101, 175, 185
commodification, 4, 8, 46, 49–50, 52–53, 62, 67, 69, 146, 158, 162–63, 188, 200
consumption, 15, 18, 24–25, 48, 69, 103, 111, 114, 129–30, 147, 149, 167, 193

DeGeneres, Ellen, 3, 8, 36, 40, 145, 185, 188, 203

Dykes to Watch Out For, 4, 19, 30, 32, 34, 53–56, 59, 167–92, 196, 200, 202

fashion
  cultural/intellectual fashions, 2, 8, 11, 19, 20, 25, 56, 135, 167, 169, 185, 191, 195
  fashion (clothing) and fashion magazines, 12, 15–16, 51, 92, 95–96, 147, 206, 207
  female masculinity, 10, 11, 31, 40, 62, 90, 92, 99, 106, 115, 117, 192
  See also masculinity
femininity, 12–16, 34–35, 46, 62, 90–93, 96–110, 112, 114–15, 119, 121, 133, 143, 152–53, 196
  See also femme
feminist
  individuals and communities, 28, 30, 38, 47, 55, 103, 161, 187, 193
  movements/activism, 19, 33, 54, 56–57, 73, 75, 98, 133, 137–39, 165, 167
  politics, 3–7, 12, 19, 48, 73, 130, 136–38, 160
  theorising, 13, 17, 20, 53, 96, 139, 154, 158, 159, 174, 175, 177
  See also lesbian-feminist
femme, 2, 51, 63, 72, 96–109, 133, 143, 146, 172–73, 192, 205
  characters, 58, 63, 75, 90–92, 111, 113, 115–16, 152, 179, 201
  theorising femmes, 11–17, 96–99, 121
film
  depictions of lesbianism, 11, 15, 25, 30, 33, 48, 61, 77, 91, 105, 110, 184, 188, 196, 203
  LGBT film festivals, 32, 148–49, 209
  pornographic film, 18, 47, 50, 70, 124, 140–62
  on television, 36, 38, 203
  theory, 138, 145, 177

gay men, 6, 24–25, 32, 38, 43, 48, 61–63, 66–68, 71, 75, 79, 81, 83–84, 114, 192, 198, 199, 204
gaze, 12, 15–17, 52, 111, 114, 141, 144–52, 155, 158, 163
genderqueer, 19, 33, 40, 117, 121, 183, 198
Gurlesque, 4, 19, 32, 50, 131, 142–44, 151, 157, 159–65, 192, 196, 208–9

Halberstam, Judith, 30, 32, 45, 115–19, 196, 210
heteronormativity, 189, 206
heterosexuality, 12, 15, 22, 52, 62, 64, 87, 91, 97–98, 100, 104–6, 116, 177, 179, 189
  heterosexual actors, 25, 106, 207
  heterosexual characters, 77, 84, 108, 209
  heterosexual men, 17, 32, 48, 63, 108, 126, 208
  heterosexual women, 35, 37, 72, 74, 92, 104, 106–8, 112, 113–14, 133, 147
homosexuality, 15, 22, 27, 29, 39–42, 46, 58, 67, 89, 105, 112, 115, 125, 189, 204

lesbian chic, 1, 3, 8, 11–13, 15–16, 18, 24–26, 35, 41, 49, 72, 90, 95, 97, 105, 133, 191, 195–96
lesbian desire, 12–14, 37, 42, 52–53, 69–75, 86, 90, 96, 103, 109, 111–12, 115–16, 121, 137, 142, 146–49, 153, 158, 207
lesbian feminist, 4, 7, 19, 38, 56, 63, 137, 161, 167, 173–74, 178–79, 184, 196
lesbian mothers, 26, 30, 36, 75–81, 86, 168, 198, 202
lesbian sex, 10, 37, 38, 42, 44, 49, 52, 66, 69–70, 86, 108, 130, 139, 140, 144–45, 157–58, 179, 199, 207, 208

Index

liberationism, 21, 23, 27, 48, 189
L Word, The, 2, 4, 7, 8, 11–13, 15, 17, 24–25, 28, 32, 34, 41, 43–46, 48, 72, 95–134, 143, 151, 165, 174, 184, 192, 195–97, 202, 205, 207–9

mainstreaming, 7, 12, 14, 21–27, 35, 141, 168, 174, 185–90, 197, 210
male gaze. See gaze
marriage, 4, 26–30, 61, 63, 66–67, 74–75, 79–85, 93, 108, 185–87, 197, 205, 207
Martin, Biddy, 4–5, 7–8, 12–13, 97–98, 105, 115, 138, 160, 195
Martindale, Kathleen, 3, 5, 19, 55–56, 135, 169, 173–74, 200
masculinity,10, 14, 15, 31–32, 37, 90, 91–92, 108, 117–22, 153, 198
  See also female masculinity
motherhood, 19, 36, 62, 66, 75–81, 99, 104, 108–9, 168, 201–2, 206
Mulvey, Laura, 17, 145, 177

Nestle, Joan, 11, 96, 139–40, 158, 164, 198

parenting. See motherhood
passing, 14, 76, 99–101, 114, 119, 122
performance
  gender performance, 6, 16, 22, 100, 109, 119, 120
  lesbian performance, 106, 147
  live performance, 29, 32, 34, 47, 50, 142–43, 147, 151, 157, 159–65, 195, 208
pornography, 12, 16–19, 44, 47–53, 63, 73, 107, 123–33, 135–65, 169, 192–93, 198, 200, 208–9

Queer as Folk, 8–9, 13, 15, 17, 25, 32, 38, 44–46, 58–59, 61–93, 98, 108, 109, 164–65, 184–85, 192, 199, 202

queer theory, 4–7, 9, 13, 16, 34–35, 139, 159, 177–82

race, 7, 22, 99, 111, 170–72, 193, 202

sex-positive, 19, 47, 50, 72, 115, 160, 163, 165, 204
sex radical, 2, 4, 6, 19, 50–51, 63, 135, 137, 148, 150, 162–63, 167, 173, 193
sex wars, 3–6, 16–17, 19, 33, 47–53, 63–64, 72–73, 96, 127, 133–36, 139–40, 148, 160, 164–65, 168–69, 174, 193, 203, 209
Showtime, 38, 41, 44, 126, 128, 132, 202, 206, 208
spectatorship, 17, 52, 141, 143–53, 160, 200

television, 1–3, 11, 16, 27, 30, 35–47, 55, 61–63, 67, 70, 82–83, 87, 93, 95–96, 108, 121, 134, 185, 191, 197–200, 203–5
transgender
  transgender and transsexual men, 30, 32–33, 114, 117–23, 183, 188, 196
  transgender and transsexual women, 32–34, 160, 183–84, 207, 208
  transgendered and transsexual people/communities, 21, 29, 33, 183, 198, 200

visibility, 14–15, 24–27, 29, 41, 45–46, 49, 52, 59, 64, 68, 77, 95, 97, 100, 133, 159, 175, 185, 188

Walker, Lisa, 11, 13, 15, 97–99, 102–3, 107, 112, 205
wedding, 27, 29–30, 38, 44, 71–72, 75, 77, 81–85, 197, 203, 205
  See also marriage
Wilton, Tamsin, 36, 48, 51–52, 136, 138–39, 140, 200